I LIBRI DE «IL COLLE DI GALILEO»

– 2 –

Enrico Fermi's
IEEE Milestone in Florence
For his Major Contribution
to Semiconductor Statistics,
1924-1926

edited by
Gianfranco Manes
Giuseppe Pelosi

Firenze University Press
2015

Enrico Fermi's IEEE Milestone in Florence : for his Major Contribution
to Semiconductor Statistics, 1924-1926 / edited by Gianfranco Manes,
Giuseppe Pelosi. – Firenze : Firenze University Press, 2015.
(I libri de «Il Colle di Galileo» ; 2)

http://digital.casalini.it/9788866558514

ISBN 978-88-6655-850-7 (print)
ISBN 978-88-6655-851-4 (online)

Graphic Design: Alberto Pizarro Fernández, Pagina Maestra snc
Back cover photo: The first operating transistor, developed at Bell Laboratories
(1947)

IEEE History Center Press
available as open access PDF and Print on Demand
http://ethw.org/Archives:IEEE_History_Center_Book_Publishing

Peer Review Process
All publications are submitted to an external refereeing process under the responsibility of the FUP Editorial
Board and the Scientific Committees of the individual series. The works published in the FUP catalogue are
evaluated and approved by the Editorial Board of the publishing house. For a more detailed description
of the refereeing process we refer to the official documents published in the online catalogue of the FUP
(www.fupress.com).

Firenze University Press Editorial Board
G. Nigro (Co-ordinator), M.T. Bartoli, M. Boddi, R. Casalbuoni, C. Ciappei, R. Del Punta, A. Dolfi, V. Fargion, S.
Ferrone, M. Garzaniti, P. Guarnieri, A. Mariani, M. Marini, A. Novelli, M. Verga, A. Zorzi.

© 2015 Firenze University Press
Università degli Studi di Firenze
Firenze University Press
Borgo Albizi, 28, 50122 Firenze, Italy
www.fupress.com/
Printed in Italy

UNIVERSITÀ
DEGLI STUDI
FIRENZE

DINFO
DIPARTIMENTO DI
INGEGNERIA
DELL'INFORMAZIONE

Index

Preface

Luigi Dei
Rector of the University of Florence

The University of Florence was born in 1924 as the transformation of some Institutes of Advanced Studies. It is interesting to notice that, among the professors of the new University, the first and second academic years we find Enrico Fermi, as teacher of Mechanics at the School of Engineering and Mathematical Physics at the School of Sciences. This circumstance has an important meaning today in relationship to the decision of dedicating to Enrico Fermi the IEEE (Institute of Electrical and Electronics Engineering) Milestone for 'his major contribution to semiconductor statistics' assigning the plaque to the School of Engineering of our University. I remember that when I was student of chemistry, statistics extraordinarily fascinated me and in the memories of nowadays four giants crowd my mind: Bose, Einstein, Fermi and Dirac. During his period in Florence Fermi worked at the statistics distribution of atom's energy in a gas. When I studied the behaviour of gas specific heats during the statistical thermodynamics course given by Giorgio Taddei in 1979-1980, I was impressed by the elegant way Fermi solved the problem with his statistical approach. Indeed, I was learning about Fermi-Dirac statistics: the Italian physicist was associated to Paul Dirac, who independently reached the same result, but starting from quantum mechanics rather than from a thermodynamic property. Some years later during my PhD course I attended some lessons on solid state physics and again the Fermi-Dirac statistics appeared on the blackboard: I was learning about theory of conduction in metals and semiconductors and again the name of Fermi appeared as the Fermi energy level. Some apparently only theoretical issues revealed themselves to be wonderfully useful for the development of the materials symbols of the digital and informatics revolution. Electronics with its transistors was born even thanks to the pioneering theoretical studies by Enrico Fermi germinated in our city and University ninety years ago. Giuseppe Pelosi *et al.* in their contribution to this book underline that the Nobel Prize in Physics 1956 John Bardeen put in the first equation of his Nobel Laureate Lecture the formula of the Fermi-Dirac distribution developed at Florence 1924-1926. It is worthwhile to mention that the motivation of this Nobel Prize was 'for the researches on semiconductors and the discovery of the transistor effect'. In the contemporary world where electronics abruptly changed forever our life and costumes, putting the world in the hands of everybody by a simple 'clic', it is fundamental to recall the role played by the scientists in determining the development and the progress of mankind. The Milestone we are celebrating today in Florence that makes Enrico Fermi in Florence close to the other two Milestones of Alessandro Volta in Pavia and Guglielmo Marconi in Sasso Marconi has to be not only an homage to a great scientist, but a declaration that science and reason must be the lights by means of which we must illuminate our future.

Foreword

Enrico Del Re
Head of the Department of Information Engineering
University of Florence

The Department of Information Engineering (DINFO) of the University of Florence is proud and honored to welcome the assignment of Enrico Fermi's IEEE Milestone to the School of Engineering 'for his Major Contribution to Semiconductor Statistics, 1924-1926'.

DINFO is the reference department of the University of Florence for all the Information and Communications Technology (ICT) fields. It carries out advanced researches in all areas of the IEEE Societies, with specific emphasis on electronics, computer engineering, telecommunications, system theory, electromagnetism, operation research, bioengineering and electrical engineering. It has many professors and researchers as IEEE Members and Senior Members and four IEEE Fellows. It also has an IEEE Student branch since '80s. DINFO research groups study and develop advanced solutions for wireless networks, signal processing, radar, remote sensing, electronic equipment, control systems, satellite systems, advanced software and cyber security systems. All these research fields share the common characteristic of dealing with the acquisition, processing, storage, transmission and utilization of information and knowledge, key resources in the present and future society. This will not be possible without the seamless advances in semiconductor technology since the invention of the first transistor in 1947 and the following development of integrated circuits. Enrico Fermi's statistics played a fundamental role in the foundation of the basic theory for the development of these modern technologies.

The Florentine science and technology tradition in electrical and electronic engineering dates back to the nineteen-twenties with the Enrico Fermi's pioneering work while professor at the University of Florence. It continued, in the ICT area, with leading-edge researchers in electronics and microwaves at Research Institute of Electromagnetic Waves (*Istituto di Ricerca sulle Onde Elettromagnetiche* – IROE) of the National Research Council (*Consiglio Nazionale delle Ricerche* – CNR) leaded just after the World War II by Nello Carrara, a fellow student of Enrico Fermi, professor at the University of Florence and the inventor of the world-wide used term of *microwaves* proposed for the first time in the IEEE publication: N. Carrara, *The detection of microwaves*, «Proceedings of the IRE», October 1932. DINFO has some Carrara's devices including the original vacuum tube used for generating the first microwaves in his earliest experiment.

DINFO is proud and conscious of the excellent scientific tradition inherited by world leading past researchers and is firmly committed to a continuing effort to advanced research in the ICT fields. The IEEE Milestone is a fundamental encouragement and support for our future activities.

Introduction

Gianfranco Manes, Giuseppe Pelosi

Enrico Fermi – Nobel Prize for Physics in 1938 – taught at the University of Florence for two academic years only (1924-1925, 1925-1926). This period is marked, scientifically speaking, by the publication of the quantum statistics which takes his name and which is the basis of semiconductor physics, and hence of modern electronics. This book is published on the occasion of the dedication, at the School of Engineering of the University of Florence, of an IEEE Milestone. The Milestone is placed in the main hall of the School of Engineering (fig. 1).

The IEEE (Institute of Electrical and Electronic Engineers) – word's largest organization for the advancement of technological innovation in the field of electrical and electronic engineering and related fields, grants IEEE Milestone, in the framework of the 'IEEE Global History Network' program, to commemorate outstanding scientific or technological achievements. In this case the Milestone commemorates the important 1926 publications of Enrico Fermi. The plaque reads (fig. 2).

Fig. 1 – Aerial view of S. Marta complex, the historical building of the School of Engineering of Florence.

We have divided this book into two parts. The first entitled *Enrico Fermi and the semiconductor electronics* and the second containing photographic reproduction of the two papers published by Enrico Fermi in the Florentine period, introducing the statistic that will take its name:

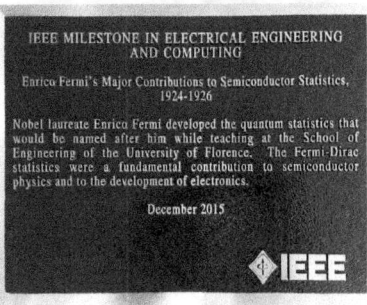

Fig. 2 – IEEE Milestone, placed at the School of Engineering of the University of Florence.

- E. Fermi, *Sulla quantizzazione del gas perfetto monoatomico*, «Atti dell'Accademia dei Lincei», vol. 3, no. 3, 1926, pp. 145-149;
- E. Fermi, *Zur Quantelung des idealen einatomigen Gases*, «Zeitschrift für Physik», vol. 36, no. 11-12, 1926, pp. 902-912.

The two parts are preceded by an introductory contribution: *The IEEE Milestones Program* by Ermanno Cardelli, Chairman of the IEEE Italy Section.

The first part of the book is divided into four contributions. The first two are re-published with the permission of the IEEE since they already appeared on IEEE publications: *Enrico Fermi in Florence* (by Giuseppe Pelosi, Massimiliano Pieraccini and Stefano Selleri) and *On the Origin of Fermi-Dirac Statistics* (by Giuseppe Pelosi and Massimiliano Pieraccini). These are followed by a contribution *From Fermi-Dirac statistics to invisible electronics* (by Gianfranco Manes), describing the evolution of electronics from the early days of vacuum tubes until today, in parallel with the evolution of communications and computer technologies. The fourth and final contribution *The role played by Fermi statistics in the evolution of Micro- and Nanoelectronics* is by Giorgio Baccarani, Alma Mater Professor at the University of Bologna.

This book is the second volume of the series of publications associated with the magazine «Il Colle di Galileo». The first volume, *Enrico Fermi a Firenze. Le lezioni di «Meccanica Razionale» al biennio propedeutico agli studi di Ingegneria: 1924-1926* [*Enrico Fermi in Florence. The lessons of «Mechanics» at the two-year preparatory courses of Engineering: 1924-1926*] (fig. 3), was also published by Florence University Press and was presented officially in March 2015 at the *Accademia dei Lincei* in Rome.

Fig. 3 – Cover of the *Enrico Fermi a Firenze* book, commemorating the years E. Fermi taught at the University of Florence and republishing his lessons of Mechanics.

The «Il Colle di Galileo» [Galileo's Hill] is a magazine describing the main activities in Physics research taking place in several Institutions in the Arcetri area, where Galileo Galilei lived. Namely: Department of Physics and Astronomy (University of Florence), Galileo Galilei Institute for Theoretical Physics (National Institute for Nuclear Physics and University of Florence), National Institute of Optics (National Research Council) and Astrophysical Observatory of Arcetri. Enrico Fermi was in Arcetri in the two years he taught at the University of Florence. In this context it is then important to remember that the European Physical Society (EPS) has designated the Arcetri hill, rich in buildings of historical and scientific interest, as an EPS Historic Site. Travelling up the hill, one can see (fig. 4):

- the headquarters of the former Institute of Physics, commissioned by Antonio Garbasso in 1921. A group of brilliant physicists, such as Gilberto Bernardini, Enrico Fermi, Giuseppe Occhialini, Giulio Racah, Franco Rasetti and Bruno Rossi, worked here in the 1920s-1930s. And it was here that Enrico Fermi wrote his fundamental work on the statistics of electrons in 1926;
- the National Institute of Optics, founded in 1927 by Vasco Ronchi, a leading light in the rebirth of optics in Italy;
- the Astrophysical Observatory of Arcetri, built in 1872 on the initiative of Giovan Battista Amici and Giovan Battista Donati. Giorgio Abetti was later to play a crucial role in its development;
- Villa Il Gioiello, lying higher up the hill just outside the complex. This is where Galileo Galilei spent the last years of his life (1631-1642) and completed writing his fundamental work entitled *Discourses and Mathematical Demonstrations Relating to Two New Sciences* (1638).

Fig. 4 – Plaque commemorating notable researchers who worked in Florence at Arcetri.

The IEEE Milestones Program

Ermanno Cardelli
Chairman of the IEEE Italy Section

Originally founded in 1884, the IEEE (Institute of Electrical and Electronic Engineers) is a non-profit professional association, the largest in the world devoted to advancing technological innovation in electrical, electronic engineering, and related fields. The association counts more than 400.000 members in more than 150 countries.

The current IEEE Italy Section has been founded the 29-th of June 2005 and we have recently celebrated its 10-th anniversary. The Italy Section is the results of the merging of the Nort Italy and Center and South Italy Sections. IEEE is active in Italy from 1959, originally as IRE Section in Milan, and is, with the IEEE Benelux, the older Section in Region 8 (Europe, Africa and Middle East). The IEEE Italy Section counts on more than 4700 members, 33 Chapters, 2 Affinity Groups and 20 Student Branches.

Since its inception, IEEE has had a standing History Committee advising the IEEE Board of Directors on matters of the legacy and heritage of IEEE and its members and their related professions and technologies, and carrying out some activities in those areas.

In 1980, in anticipation of its Centennial celebration in 1984, IEEE established the IEEE History Center to be the staff arm of the History Committee. In 1990, the Center moved to the campus of Rutgers University, which became a cosponsor. In 2010, the Center also entered into a cooperative agreement with the University of California, Merced, in order to have a presence near the high-tech centers of the US's west coast. In 2014, the Center moved to Stevens Institute of Technology. Today, IEEE's central historical activities are carried out largely by the staff of the History Center, under the guidance of the History Committee.

The mission of the IEEE History Center is to preserve, research and promote the history of information and electrical technologies. The Center maintains many useful resources for the engineer, for the historian of technology, and for anyone interested in the development of electrical and computer engineering and their role in modern society. The History Center building is not a museum, and does not contain artifacts or exhibits, being merely offices and the library. Visiting scholars and researchers are welcome to use our research library and archives, by appointment only. To make an appointment, please contact ieee-history@ieee.org.

Most of the Center's resources are available online at the Engineering and Technology History Wiki. The Center's holdings include the IEEE Archives, which consist of the unpublished records of IEEE and a collection of historical photographs relating to the history of electrical and computer technologies, and a collection of oral history transcripts of pioneering engineers. Within IEEE, the History Center is part of IEEE Corporate Activities, and it is co-sponsored by Stevens Institute of Technology, where it is affiliated with the College of Arts and Letters.

The History Committee, a Committee of the IEEE Board of Directors, is responsible for promoting the collection, writing, and dissemination of historical information in the fields covered by IEEE technical and professional activities, as well as historical information about the IEEE and its predecessor organizations. It provides assistance to all major organizational units, works with institutions of a public nature such as the Smithsonian Institution when helpful information is requested and can be secured, and provides information and recommendations to the IEEE Board of Directors when appropriate.

Typically, the History Committee holds two face-to-face meetings per year, with additional teleconferences scheduled as needed. The face-to-face meetings are usually in early March and early November of each year. Current History Committee Members are:

David E. Burger, Chair, Australia
Allison Marsh, South Carolina, USA
Fiorenza Albert-Howard, Canada
Jacob Baal-Schem, Israel
Theodor Bickart, Colorado, USA
Richard Gowen, South Dakota, USA
John Impagliazzo, New York, USA
Paul Israel, Milestone coordinator for Regions 1-7, New Jersey, USA
David Kemp, Canada
Antonio Perez-Juste, Spain
Antonio Savini, Milestone coordinator for Region 8-10, Italy
Mischa Schwartz, New York, USA
Isao Shirakawa, Japan
Jorge Soares, Switzerland
John Vig, New Jersey, USA

The IEEE Milestones in Electrical Engineering and Computing program honors significant technical achievements in all areas associated with IEEE. It is a program of the IEEE History Committee, administered through the IEEE History Center. Milestones recognize the technological innovation and excellence for the benefit of humanity found in unique products, services, seminal papers and patents. Milestones are proposed by any IEEE member, and are sponsored by an IEEE Organizational Unit (OU) – such as an IEEE section, society, chapter or student branch. After recommendation by the IEEE History Committee and approval by the IEEE Board of Directors, a bronze plaque commemorating the achievement is placed at an appropriate site with an accompanying dedication ceremony.

IEEE established the Milestones Program in 1983 in conjunction with the 1984 Centennial Celebration to recognize the achievements of the Century of Giants who formed the profession and technologies represented by IEEE.

Each Milestone recognizes a significant technical achievement that occurred at least twenty-five years ago in an area of technology represented in IEEE and having at least regional impact. To date, more than a hundred Milestones have been approved and dedicated around the world.

The list of Milestones in Region 8 follows

Year	Event	IEEE Section
1994	Poulsen Arc Radio Transmitter	Denmark
1999	**Volta's Battery**	**Italy**
1999	Operational Use of Wireless	South Africa
2000	County Kerry Wireless Station	United Kingdom and Ireland
2001	Transmission of Transatlantic Radio	United Kingdom and Ireland
2002	Shannon Electrification	United Kingdom and Ireland
2002	Transatlantic TV by Satellite	France, United Kingdom and Ireland
2002	Pioneering Work on Quartz Watch	Switzerland
2003	Marconi Early Wireless Experiments	Switzerland
2003	Franklin's Work in London	United Kingdom and Ireland
2003	Bletchley Park Codebreaking	United Kingdom and Ireland
2004	Fleming Valve	United Kingdom and Ireland
2004	Lempel-Ziv Algorithm	Israel
2005	Popov's Radio Work	Russia Northwest
2005	Vucje Hydroelectric Plant	Yugoslavia
2005	CERN Instrumentation	France
2006	Callan's Pioneering Contributions	United Kingdom and Ireland
2006	WEIZAC Computer	Israel
2006	TAT-1 Telephone Cable	United Kingdom and Ireland
2007	Early Remote Control	Spain
2009	Maxwell's Equations	United Kingdom and Ireland
2009	Shilling Early Telegraph	Russia Northwest
2009	Compact Disc Player	Benelux
2010	Star of Laufenburg Interconnection	Switzerland
2010	Branly's Discovery of Radioconduction	France
2010	Public-Key Cryptography	United Kingdom and Ireland
2011	Discovery of Superconductivity	Benelux
2011	**Marconi's First Wireless Experiments**	**Italy**
2013	Krka-Sibenik Power System	Croatia
2013	Invention of Holography	United Kingdom and Ireland
2014	First Breaking of Enigma	Poland
2014	Rheinfelden Hydroelectric Plant	Germany
2014	Hertz Proof of Electromagnetic Waves	Germany
2015	Invention of Stereo Sound Reproduction, 1931	United Kingdom and Ireland

As reported in the table above, in Italy were placed two Milestones, up to now. The first placed on September 1999 in Como (Italy) and dedicated to Alessandro Volta's invention of the electrical battery (1799) (fig. 1). The second placed on April, 2011 in Sasso Marconi (Bologna, Italy) commemorating Guglielmo Marconi's first experiments of wireless telegraphy (1894-1895) (fig. 2).

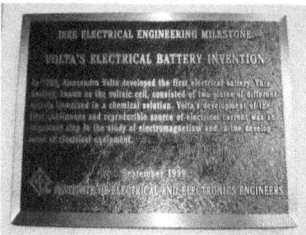

Fig. 1 – A photo of the IEEE Milestone placed at the Electric Technique Museum of the University of Pavia, Italy and its text.

Volta's Electrical Battery Invention, 1799

Como, Italy, Dedicated September 1999 – IEEE North Italy Section

In 1799, Alessandro Volta developed the first electrical battery. This battery, known as the Voltaic Cell, consisted of two plates of different metals immersed in a chemical solution. Volta's development of the first continuous and reproducible source of electrical current was an important step in the study of electromagnetism and in the development of electrical equipment.

Fig. 2 – The plaque next to Villa Griffone (Sasso Marconi, Bologna), in front of the Celestini hill in a moment of the inauguration ceremony and the Milestone text.

Marconi's Early Experiments in Wireless Telegraphy, 1895

Pontechio Marconi, Italy, Dedicated 29 April 2011 -- IEEE Italy Section

In this garden, after the experiments carried out between 1894 and 1895 in the 'Silkworm Room' in the attic of Villa Griffone, Guglielmo Marconi connected a grounded antenna to its transmitter. With this apparatus the young inventor was able to transmit radiotelegraphic signals beyond a physical obstacle, the Celestini hill, at a distance of about two kilometres. The experiment heralded the birth of the era of wireless communication. On this hill, during the summer of 1895, the radiotelegraphic signals sent by Guglielmo Marconi from the garden of Villa Griffone were received. The reception was communicated to Marconi with a gunshot. This event marked the beginning of the new era of wireless communication.

Part I
Enrico Fermi
and Semiconductor Electronics

Enrico Fermi in Florence

Giuseppe Pelosi, Massimiliano Pieraccini, Stefano Selleri

A Brief Biography (from 1921 to 1926)

Enrico Fermi (Rome [Italy], September 29, 1901-Chicago [Illinois, USA], November 28, 1954) got his degree in Physics in 1921 at the *Scuola Normale Superiore* in Pisa (Italy). Among his fellow students Fermi had Nello Carrara (Florence [Italy], February 19, 1900-Florence [Italy], June 5, 1993), who introduced the term *microwaves* in scientific literature and with whom the readers of this Magazine have a long term acquaintance[1,2] (fig. 1). Another personality known to our readers is Gian Antonio Maggi (Milan [Italy], February 19, 1856-Milan [Italy], July 12, 1937) who was among his teachers and on whom was dedicated a former Historical Corner[3].

Just after graduation Fermi spent a semester at the University of Göttingen (Germany), studying under Max Born and where he met also Werner Heisenberg. He was then in Leiden (The Netherlands) with Paul Ehrenfest, meeting also Hendrik Lorentz and Albert Einstein.

Before becoming the prominent scientist we know, Enrico Fermi came back to Italy and spent a couple of years in Florence (fig. 2). This is seldom remembered and scarcely documented, yet it is relevant for the electronic community. The only records of this period are in Emilio Segrè book (one of Fermi's students, and himself Nobel Laureate in Physics in 1959)[4] and in the book by Enrico Fermi's wife Laura Fermi[5].

Fermi joined the Arcetri Institute of Physics of the University of Florence, which was directed at that time by Antonio Garbasso (Vercelli [Italy], April 16, 1871-Florence [Italy], March 14, 1933) (fig. 3), renowned physicist, which studied with H. Hertz in Bonn and H. von Helmholtz in Berlin. He was also an active politician, being Major of Florence from 1920 to 1928 and senator of the Kingdom of Italy since 1926. Garbasso was indeed more in politics than in science when he called Fermi to Arcetri[6].

© 2014 IEEE. Reprinted, with permission from «IEEE Antennas and Propagation Magazine», vol. 55, no. 6, 2013, pp. 272-276.

[1] G. Pelosi, *The birth of the term microwaves*, «Proceedings of IEEE», vol. 84, no. 2, 1996, p. 326.

[2] G. Abbatangelo, M. Cavicchi, G. Manara, G. Pelosi, P. Quattrone, S. Selleri, *The early history of radio communications through the photos of the Italian Navy archives*, «IEEE Antennas and Propagation Magazine», vol. 48, no. 6, 2006, pp. 224-236.

[3] O.M. Bucci, G. Pelosi, *From wave theory to ray optics*, «IEEE Antennas and Propagation Magazine», vol. 36, no. 4, 1994, pp. 35-42.

[4] E. Segrè, *Enrico Fermi, Physicist*, University of Chicago Press, Chicago, IL, 1st edition, 1972.

[5] L. Fermi, *Atoms in the Family. My life with Enrico Fermi*, University of Chicago Press, Chicago, IL, 1st edition, 1954.

[6] W. Joffrain, *Un inedito di Enrico Fermi, Elettrodinamica [An unpublished manuscript by Enrico Fermi: Electrodynamics]*, XVII Conference on Physics and Astronomy History [in Italian], Como, Italy, May 15-19, 1998, pp 1-11.

Fig. 1 – Historical photograph taken on the Apuane Alps in 1920. From left to right, Enrico Fermi, Nello Carrara and Franco Rasetti (Pozzuolo Umbro [Italy], August 10, 1901-Waremme [Belgium], December 5, 2001) physicist, botanist and paleontologist. Another rare photograph of young Enrico Fermi with Nello Carrara was published on this Magazine's Historical Corner[7].

Fig. 2 – (*top*) Enrico Fermi's signature in the Member's book of the Gabinetto Scientifico-Letterario G.P. Vieusseux, a very important cultural association in Florence, founded at the beginning of the XIX century. Next to the signature there is Fermi's home address in Florence, reading «Via Pian de Giullari 63/a». Address is very close to the house of Galileo Galilei at Arcetri (*bottom*), now belonging to the University of Florence.

In his few years in Florence Enrico Fermi undertook both educational and research activities, these latter having a huge, but somewhat underestimated, impact on electronics. These achievements will be briefly detailed in the following.

In 1926, he then applied for a professorship at the "Sapienza" University of Rome, one of the first three chairs for theoretical physics in Italy, created at the urging of Professor Orso

[7] P. Mazzinghi, G. Pelosi, *Enrico Fermi talks about Guglielmo Marconi*, «IEEE Antennas and Propagation Magazine», vol. APM-53, no. 3, 2011, pp. 226-230.

Mario Corbino (Augusta [Italy], April 30, 1876-Rome [Italy], January 23, 1937) (fig. 4), chair of experimental physics, Director of the Institute of Physics, and member of Benito Mussolini's cabinet. Corbino then helped Fermi in recruiting his new team, among others: Edoardo Amaldi, Bruno Pontecorvo, Ettore Majorana, Emilio Segrè and Franco Rasetti, soon known as the 'Via Panisperna boys' after the street where the Institute of Physics was located (fig. 5).

Fig. 3 – Antonio Garbasso, Director of the Arcetri Institute of Physics, about 1925.

Fig. 4 – Orso Mario Corbino, chair of Experimental Physics at the "Sapienza" University of Rome when Enrico Fermi won the chair of Theoretical Physics.

Fig. 5 – The Institute Physics Institute on Via Panisperna, Rome.

Educational Activities in Florence

Enrico Fermi taught Mechanics (*Meccanica Razionale*) – at the School of Engineering and at the Faculty of Sciences – and Mathematical Physics – at the Faculty of Sciences – as reported in several documents of the University of Florence.

The course of Mechanics lessons were picked up by two Fermi's students and their handwritten notes of the course *Meccanica Razionale* were printed[8], a rare copy of this book is conserved at the Temple University (Philadelphia, Pennsylvania, USA) (fig. 6).

Fig. 6 – Front cover of the lesson notes of the Mechanics (Meccanica Razionale) course of Enrico Fermi, at the School of Engineering of the University of Florence, taken by the students B. Bonanni and P. Pasca. [Courtesy of Temple University (Philadelphia, Pennsylvania, USA)][9]

Even more interesting for the reader are the lectures on Electrodynamics Fermi gave in the same period, within the Mathematical Physics, whose notes were gathered as a manuscript, available in three copies.

The first copy, of reduced length, kept in Pisa, is probably not written by Fermi himself, as the many citations to other Fermi's works suggests[10] and is probably the latest of the three versions; a second copy, much longer and kept in Rome has been credited to Fermi's student A. Morelli, of Rome and covers the lectures given by Fermi in Rome in 1926, which strictly follows those given in Florence in 1924 and 1925. The original notes are probably those kept at the University of Chicago (Illinois, USA), donated by Fermi's widow Laura Fermi in 1955 and which have been dated to 1925 (fig. 7)[11].

[8] E. Fermi, *Lezioni di Meccanica Razionale*, Litografia L. Tassini, Florence (Italy), 304 pages [in Italian]. The year of publication (1926) can only be presumed by the handwritten dedication on the front cover.
[9] Fermi, *Lezioni di Meccanica Razionale*, op. cit.
[10] Joffrain, *Un inedito di Enrico Fermi, Elettrodinamica*, op. cit.
[11] Joffrain, *Un inedito di Enrico Fermi, Elettrodinamica*, op. cit.

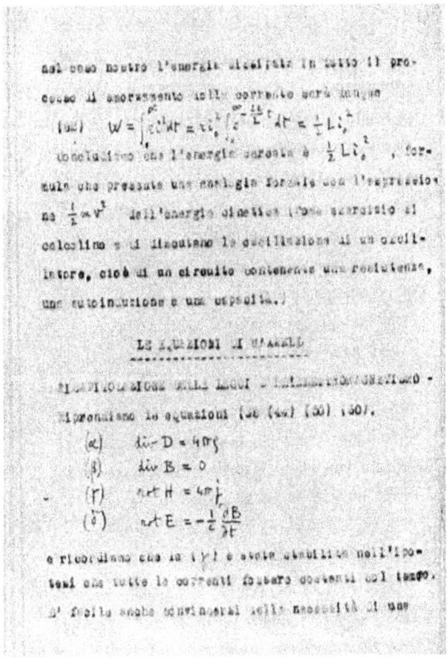

Fig. 7 – First page of Maxwell's equation chapter from Fermi's typescript (page 50)[12]. According to W. Joffrain[13,14] this is the original copy of the notes, possibly with Fermi's handwriting.

It is important to mention that the Electrodynamics notes were re-published after a critical investigation on the three manuscript in 2006 and are now available (in Italian)[15].

The Mechanics (*Meccanica Razionale*) lecture notes, in the version kept at the Temple University Library, are now being processed by the Firenze University Press and an edition is foreseen for next year.

Research Activities in Florence

During his period in Florence Fermi worked and published, early in 1926, one of his major scientific contributions[16,17]: the calculus of the statistic distribution of the atoms' energy in a gas. This was a fundamental question at that time. The classical mechanics applied to atoms gave results for the specific heat not in agreement with measurements. Fermi's statistic distribution solved the question, as did in the October of the same year Paul Dirac, publishing the same results[18]. Fermi wrote a letter to Dirac to claim his priority and the English scientist kindly recognized it[19].

[12] E. Fermi, *Elettrodinamica [Electrodynamics]* [In Italian, with English preface], edited by W. Joffrain, U. Hoepli, Milan (Italy), 2006.

[13] Joffrain, *Un inedito di Enrico Fermi, Elettrodinamica*, op. cit.

[14] Fermi, *Elettrodinamica*, op. cit.

[15] Fermi, *Elettrodinamica*, op. cit.

[16] E. Fermi, *Sulla quantizzazione del gas perfetto monoatomico [On the quantization of an ideal monoatomic gas]*, «Atti dell'Accademia dei Lincei», vol. 3, no. 3, 1926, pp. 145-149.

[17] E. Fermi, *Zur Quantelung des idealen einatomigen Gases*, «Zeitschrift für Physik», vol. 36, no. 11-12, 1926, pp. 902-912.

[18] P.A.M. Dirac, *On the theory of quantum mechanics*, «Proceedings of the Royal Society of London», vol. A112, 1926, pp. 281-305.

[19] G. Farmelo, *The Strangest Man: the Hidden Life of Paul Dirac, Quantum Genius*, Faber & Faber, London, UK, 2009.

Giuseppe Pelosi, Massimiliano Pieraccini, Stefano Selleri

The Fermi-Dirac distribution, as it was known, refers to a gas, but has more general validity. The first to apply it to semiconductors was Alan Wilson in 1931[20]. Finally The Nobel Prize in Physics 1956 was awarded jointly to William B. Shockley, John Bardeen and Walter H. Brattain «for their researches on semiconductors and their discovery of the transistor effect» (fig. 8). Significantly the first formula that Bardeen (fig. 9) showed at the this Nobel Lecture was the Fermi-Dirac distribution[21,22].

Fig. 8 – (from left to right) John Bardeen (Madison [Wisconsin, USA], May 23, 1908-Boston [Massachusetts, USA], January 30, 1991), Walter H. Brattain (Amoy [China], February 10, 1902-Seattle [Washington, USA], October 13, 1987) and William B. Shockley (London [United Kingdom], February 13, 1910-Stanford [California, USA], August 12, 1989), winners of 1956 Nobel Prize for Physics for their work on the transistor.

JOHN BARDEEN

Semiconductor research leading to the point contact transistor

Nobel Lecture, December 11, 1956

[...]

Occupancy of the levels is given by the position of the Fermi level, E_F. The probability, f, that a level of energy E is occupied by an electron is given by the Fermi-Dirac function:

$$f = \frac{1}{1 + \exp{(E - E_F)}/kT}$$

Fig. 9 – Title and first equation of the Nobel Lecture by John Bardeen (1956). Bardeen is the only one to have ever won two Nobel Prizes in Physics, this in 1956 (for the transistor), the second in 1972 (for superconductivity).

Acknowledgments

The authors would like to thank Prof. Alberto Tesi, Rector of the University of Florence, for having granted access to the University archives, and Prof. Roberto Vergara Caffarelli, of the University of Pisa, for having provided archive material.

[20] A.H. Wilson, *The theory of electronic semi-conductors*, «Proceedings of the Royal Society of London», vol. A133, 1931, pp. 661-677.
[21] G Busch, *Early history of the physics and chemistry of semiconductors-from doubts to fact in a hundred years*, «European Journal of Physics», vol. 10, no. 4, 1989, p. 254.
[22] J. Berdeen Nobel Lecture 1956, http://www.nobelprize.org/nobel_prizes/physics/ laureates/1956/bardeen-lecture.pdf (11/15).

On the origin of Fermi-Dirac Statistics

Giuseppe Pelosi, Massimiliano Pieraccini

The Italian-American Nobel Prize Enrico Fermi (Rome, Italy, 1901-Chicago, Illinois, USA, 1954) is universally known for the so-called 'Fermi-Dirac statistics'[1] that is the basis of the theory of conduction in metals and semiconductors, but not everybody knows how, when and where he conceived this fundamental contribution to the modern electronics. Indeed, as we show in this contribution, his mind was addressed to the theoretical issues that would lead him to 'his' statistics since he was a young graduate, in Florence.

Significantly, his graduation thesis for the *Scuola Normale Superiore* (a public learning institution in Pisa renowned for its excellence), on June 22, 1922 was entitled: *Un teorema di calcolo delle probabilità ed alcune sue applicazioni* (A theorem on probability calculus and some of its applications)[2]. A year after he demonstrated another theorem about the ergodicity of the normal systems[3], but the key step forward to the statistics named after him could be retrieved in a paper he published in December 1924, in the Italian journal «Il Nuovo Cimento» (fig. 1).

He discussed the Wilson-Sommerfeld quantization rules[4,5] for the case of a set of identical systems. He makes a simple and enlightening example. Let consider three electrons that can be disposed on a ring at the vertices of an equilateral triangle. If the electrons are distinguishable, only a 360° rotation could bring the system in the initial condition, but if the electrons are undistinguishable any 120° rotation return the system in the initial condition (fig. 2). Fermi noted that the application of quantization rules gives different results for these two cases.

Therefore, since 1924 Fermi knew how to quantize a set of particles or atoms, but not how to dispose them in the energy levels. The missing piece of the puzzle was what today we call 'Pauli Exclusion Principle'. Many years after, when Fermi was in America, he regretted

Originally published in July 2014 by the authors under Creative Common "Attribution, NonCommercial-ShareAlike" license on the IEEE Engineering and Technologically History Wiki, http://ethw.org/On_the_origin_of_Fermi-Dirac_statistics

[1] C. Bernardini, L. Bonolis, *Enrico Fermi: his work and legacy*, Springer, Berlin-Heidelberg-New York, 2001.
[2] E. Fermi, *Note e Memorie [Collected Papers]*, Accademia Nazionale dei Lincei and The University of Chicago Press, 2 Volumes, Roma and Chicago, Italy and USA,1961-1965.
[3] E. Fermi, *Dimostrazione che in generale un sistema meccanico normale è quasi ergodico*, «Il Nuovo Cimento», vol. 25, 1923, pp. 267-269.
[4] W. Wilson, *The quantum theory of radiation and line spectra*, «Philosophical Magazine», vol. 29, 1915, pp. 795-802.
[5] A. Sommerfeld, *Zur Quantentheorie der Spektrallinien*, «Annalen der Physik», vol. 356, no. 17, 1916, pp. 1-94.

with his friend Emilio Segrè that, if he had had a few more months he could get it[6]. Anyway, the point is when he was aware of it. The article of Wolfang Pauli that introduced the Exclusion Principle was published in April 1925[7], but at that time Fermi was engaged in experimental works of spectroscopy[8], so probably he did not read it. In July he was member of the commission of the 'state exam', the final examination of the high school students, a hard commitment that surely distracted him from his research as he complains in his letters[9]. In August he went on vacation to the Dolomiti mountains with several friends, including the physicist Ralph de Laer Kronig. And the German scientist was probably the contact point, the stark, between Pauli and Fermi. Kronig wrote regularly to Pauli, so he was surely aware of his recent publication.

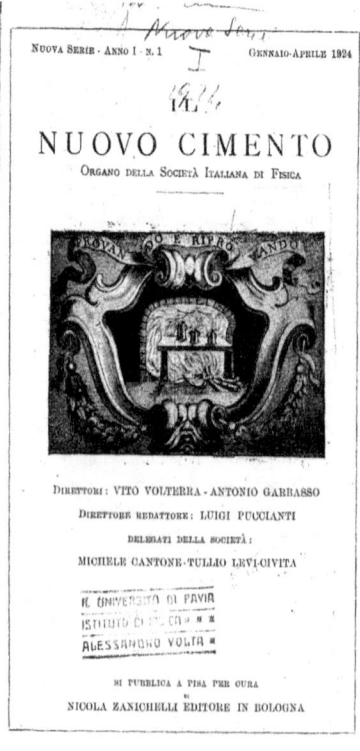

Fig. 1 – Cover the Italian scientific journal where Fermi published a preliminary work[10] on quantum statistics.

---

[6] E. Segrè, *Enrico Fermi Physicist*, University of Chicago Press, Chicago, USA, 1970.

[7] W. Pauli, *Uber den Zusammenhang des Abschlusses der Elektronengruppen im atom mit der Komplexstruktur der Spektren*, «Zeitschrift für Physik», vol. 31, 1925, pp. 765-783.

[8] Fermi, *Note e Memorie (Collected Papers)*, op. cit.

[9] Segrè, *Enrico Fermi Physicist*, op. cit.

[10] E. Fermi, *Considerazioni sulla quantizzazione dei sistemi che contengono degli elementi identici*, «Il Nuovo Cimento», vol. 1, no. 1, 1924, pp. 145-152.

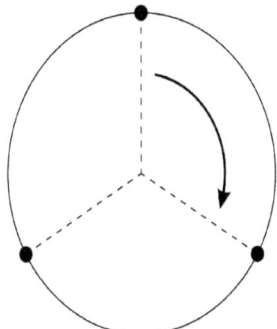

Fig. 2 – Ideal system of three electrons at the vertexes of an equilateral triangle.

On the other hand, the first public presentation of the statistical calculus of Fermi was on February 7, 1926[11] at the *Accademia dei Lincei* (the Italian Scientific Academy), therefore the gestation of the idea can be located between summer 1925 and the end of the year. In this span of time Fermi taught 'Mathematical Physics' and 'Mechanics'[12] at the University of Florence, Italy to a «number infinitely small of students» as he wrote in a personal letter[13].

He published his major result before in the «Atti dell'Accademia dei Lincei»[14] and after in the German journal «Zeitschrift fur Physik»[15] with more details. The key formula he derived is:

$$N_s = \frac{Q_s}{Ae^{\beta E} + 1}$$

with Q degeneration of quantum state s, N_s number of atoms in state s, A and β constants. This expression is a bit different from what we are accustomed to use:

$$f = \frac{1}{1 + e^{\frac{E - E_F}{kT}}}$$

with f probability of occupation of a state at energy E, E_F constant (the so-called Fermi's energy), k Boltzmann's constant, T temperature.

In effect the two articles of Fermi did not have the resonance they deserved, the community of quantum scientists understood its value only after Dirac obtained it independently form the symmetry properties of the state function that describes a set of particles[16]. So, the expression of Fermi distribution we currently use is effectively Dirac's formulation.

---

[11] E. Fermi, *Sulla quantizzazione del gas perfetto monoatomico*, «Atti dell'Accademia dei Lincei», vol. 3, no. 3, 1926, pp. 145-149.

[12] G. Pelosi, M. Pieraccini, S. Selleri, *Enrico Fermi in Florence*, «IEEE Antennas and Propagation Magazine», vol. 55, no. 6, 2013, pp. 272-276 (reprinted in this book).

[13] Segrè, *Enrico Fermi Physicist*, op. cit.

[14] Fermi, *Sulla quantizzazione del gas perfetto monoatomico*, op. cit.

[15] E. Fermi, *Zur Quantelung des idealen einatomigen Gases*, «Zeitschrift für Physik», vol. 36, no. 11-12, 1926, pp. 902-912.

[16] P.A.M. Dirac, *On the theory of quantum mechanics*, «Proceedings of the Royal Society of London», vol. A112, 1926, pp. 281-305.

From Fermi-Dirac Statistics
to the Invisible Electronics

Gianfranco Manes

> The reasonable man adapts himself to the world: the unreasonable one persists in trying to adapt the world to himself. Therefore all progress depends on the unreasonable man.
>
> [G.B. Shaw, Nobel laureate in literature in 1925]

Foreword: the Invisible Electronics

The invisible Electronics denotes a feature of modern technology. As electronic devices scale down, they become more connected and more integrated into our environment; eventually, the technology disappears into our surroundings until only the user interface remains perceivable by users, the so called *human centric* interaction. An example of this vision in represented by the Internet of Things, an easy, natural way using information and intelligence that is hidden in the network connecting the devices embedded in the environment.

The first electronic components were bulky, relatively expensive and capable of performing a few elementary functions. The electronics of today is made of micro and nano-components incorporated, or *embedded*, in everyday objects; they can transmit information at high speed, allowing anywhere, always-on connectivity, ultimately resulting in devices that dissolve in the body or within the environment. Enabling technologies are, among many others, the wireless sensor networks, allowing for remotely monitoring the main environmental parameters with at unprecedented time/space scale, and modern computers, able to develop computations with speed, size and costs that were unpredictable a few years ago. The modern electronic systems have found useful applications in places once unthinkable, from museums to provide visitors with timely information and localized, to crops, to monitor environmental conditions and the risk of pathogens, to name just two examples among the myriad of other that may be mentioned. The application of modern Electronics have pervaded our world in an often invisible and still essential and indispensable part of our every day lives.

While electronics has made extraordinary progresses since Enrico Fermi, it is also true that its foundations are still the same we taught to several generations of students, now excellent professionals and researchers. In fact, despite the whirlwind of reforms that the university has undergone in recent years, the preparation of our engineers is nevertheless remained at a good standard and internationally acknowledgement. The Italian school, at least in most of the universities, has maintained in the years the rigor and seriousness that has always distinguished, without slipping into easy compromise that perhaps, momentarily, would meet the demands of the labor market, but that in the long term would go against the best interests of the students and the companies, because it is the solid foundation that makes a difference in a professional career. Therefore, as far as we prepare *unreasonable* and imaginative professionals in our Universities, we will still fulfil our academic commitment.

Lectio Magistralis, December 4, 2015.

Quantum Mechanics and Solid State Physics

Since the beginning of XX century to the end of '30s, when the final mathematical formulation of the quantum theory was developed, the impressive evolution of Quantum Physics gave us a theory of matter and fields, that would deeply impact on radio technology driving the transition from vacuum tubes to solid state devices, eventually resulting in what we call today Electronics. Still in the XXI century, quantum mechanics will continue to provide fundamental concepts and essential tools for all the sciences.

The clue that triggered the quantum revolution came not from studies of matter but from a problem in radiation. The specific challenge was to understand the spectrum of light emitted by hot bodies: the blackbody radiation. In 1900 Max Planck gave a solution to the black-body radiation problem introducing his radiation law and Albert Einstein in 1905 developed a quantum-based theory to explain the photoelectric effect. That early formulation was significantly reformulated in the mid-1920s, which resulted in a comprehensive formulation of the quantum mechanics.

The first step was originated by some inconsistencies on the nature of light. The dual nature of light, particle-like or wave-like, had to be a theoretical enigma that had to thrill the scientists for at least 20 years.

The second step was originated by an incongruence about the structure of the atom. Following the models developed by Ernest Rutherford, it was known that atoms contain positively and negatively charged particles. According to electromagnetic theory, oppositely charged particles attract reciprocally, thus resulting in collapse of electron in the nucleus, while radiating light in a broad spectrum.

The second incongruence was partially solved by Niels Bohr in 1913. Bohr proposed a radical hypothesis: electrons in an atom exist only in certain stationary states, including a ground state. Bohr proposed a model of the atom, and while he did not solve the problem of atomic stability, his theory provided a quantitative description of the spectrum of the hydrogen atom.

In 1923 Louis de Broglie, in his Ph.D. thesis, proposed that the particle behaviour of light should have its counterpart in the wave behaviour of particles. He predicted that the wavelengths are given by Planck's constant h divided by the momentum of the $mv = p$ of the electron, i.e. $\lambda = h\ /\ mv = h\ /\ p$. The De Broglie theory did not explain the particle's wave nature but was of paramount importance for the subsequent steps .

In 1924 Satyendra Nath Bose proposed a totally new way to explain the Planck radiation law. He treated light as if it were a gas of massless particles, the photons, that do not obey the classical laws of Boltzmann statistics but behave according to a new type of statistics based on particles' indistinguishable nature. Einstein immediately applied Bose's reasoning to a real gas of massive particles and obtained a new law, to become known as the Bose-Einstein distribution.

In the three-year period from January 1925 to January 1928 a number of critical events happened, that revolutionized the foundation of classic physics and laid down the foundation of a new science, the Quantum Mechanics.

It is, however, interesting to note that the principal actors in the creation of quantum theory were young. In 1925 Wolfgang Pauli was 25 years old, Heisenberg and Enrico Fermi were 24, and Paul Dirac and Pascual Jordan were 23. Werner Heisenberg, was 26 years old when he laid the foundations for atomic structure theory by obtaining an approximate solution to Erwin Schrödinger's equation for the helium atom in 1927, and general techniques for calculating the structures of atoms were created. Schrödinger, at age 36, was a *late bloomer*. Max Born and Bohr were even older, and it is significant that their contributions were largely interpretative.

The Fermi's Statistics and its impact on the Physics of Semiconductors[1]

The Fermi statistics along with Felix Bloch's band theory, whose important parameter is the Fermi level, played a major role in the understanding of crystal lattice band structure and the allowed energy levels, an essential step for the understanding of semiconductors and, then, of all solid-state devices.

The band theory, in particular, is the foundation of the modern theory of solids, as it led to understanding of the nature and explained the important properties of semiconductors. The Fermi function $f(E)$ gives the probability that a given available electron energy state will be occupied at a given temperature, while the position of the Fermi level located halfway of the forbidden band, the energy gap between the valence and conduction bands, corresponding to the energy with 50% probability of occupation.

$$f(E) = \frac{1}{e^{(E-E_F)/kT} + 1}$$

The width of the 'forbidden' band is the key variable in the band theory; it defines the electrical and optical properties of the material. For example, in semiconductors, the conductivity can be increased by creating an allowed energy level in the band gap by doping, i.e. introducing additives into the base material to alter its physical and chemical properties.

The transition of an electron from the valence band into the conduction band is called the charge carrier generation process (carriers of a negative charge are electrons and carriers of a positive charge are holes), and the reverse transition is called the recombination process.

The increase in conductivity with temperature in semiconductors can be modelled in terms of the Fermi function, which allows one to calculate the population of the conduction band.

Figure 1 shows the implications of the Fermi function for the electrical conductivity of a semiconductor. According to Fermi statistics and taking in account for the distribution of electrons and holes and their respective densities-of-states, it is possible to determine the density of the electrons, n, in the conduction band and the density of the holes, p, in the valence band at any spatial point in the semiconductor once the densities-of-states and the Fermi level at that point are known.

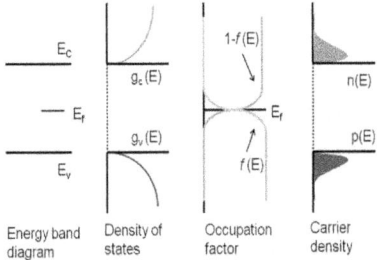

Fig. 1 – Carrier density in a semiconductor derived using band theory, state density and Fermi statistics.

---

[1] This subject is discussed in greater details by Professor G. Baccarani in another chapter of the book.

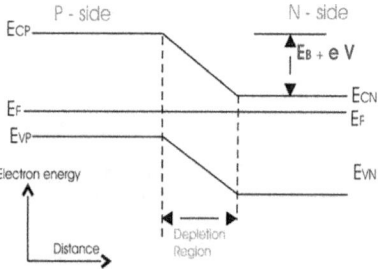

Fig. 2 – Schematic diagram of *p-n* junction explained in terms of band alignment.

Those values of concentration are needed to formulate the drift-diffusion equation, one of the simplest and most popular semiconductor models, derived by Willy Werner Van Roosbroeck in 1950.

$$J_n(x) = q\mu_n n(x)\varepsilon(x) + qD_n \frac{dn(x)}{dx} + q\mu_p p(x)\varepsilon(x) + qD_p \frac{dp(x)}{dx}$$

The drift diffusion model, here represented in monodimensional form, plays a key role in the understanding and modeling the operation of *p-n* junction (fig. 2).

The birth of radio: from cable to wireless

The birth of radio communication can be virtually located in 1864 when James Clerk Maxwell first proved the existence electromagnetic waves and presented his results to the Royal Society. The four differential equations describing the associated electric and magnetic field are universally known as Maxwell's Equations.

Following that, Heinrich Hertz, in a series of experiments started in 1887, proved the physical existence of radio waves that Maxwell had shown to exist mathematically, using a simple spark gap across an induction coil with a loop of wire to act as an antenna. The receiver consisted of a smaller gap in a loop having the same size as that in the transmitter. In his experiments Hertz also discovered many of their properties.

In the autumn of 1894 Guglielmo Marconi did the first experiments with radio waves, only achieving distances of a few metres, eventually managing to send signals over a distance of about 2 kilometres.

Since then, new experiments came in rapid succession, paving the way to a dramatic development of radio techniques and technologies. On 13 May 1897, Marconi set up a link spanning the 14 kilometres of the Bristol Channel and sent the world's first ever wireless communication over open sea and in subsequent experiments demonstrated the feasibility of propagation over water.

After this Marconi put on many other demonstrations and gave lectures and in this way he was able to gain the maximum amount of publicity, which stimulated the interest of other experimenters[2].

---

[2] In 1909 Guglielmo Marconi was awarded the Nobel Prize Physics jointly with Karl Ferdinand Braun 'in recognition of their contributions to the development of wireless telegraphy'.

In the years 1921-23, transatlantic communications succeeded using short wave. Until then, long distance communications had been concentrated on the long wavelengths.

The Early Radio Receivers and the Era of Vacuum Tubes

For his early experiments Marconi had used very simple and primitive equipment. A first improvement was represented by the, crystal rectifier whose properties were discovered in 1874 by Karl Ferdinand Braun[3]. Those crystal detectors were originally made from a piece of crystalline mineral such as galena or silicon intimately contacted to a metal wire and often referred to as cat's whisker detector, see figure 3.

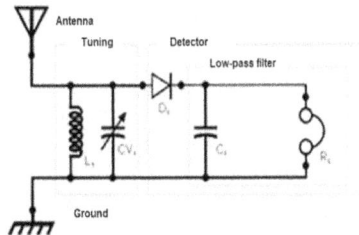

Fig. 3 – Crystal radio receiver. (*top*) Radio receiver schematic diagram. (*bottom*) A radio receiver set on the left the cat's whisker detector is clearly visible.

The use of the metal-semiconductor diode rectifier was proposed by Julius Edgar Lilienfeld in 1925 in the first of his three transistor patents as the gate of the metal-semiconductor field effect transistors.

However, the fabrication of those early solid-state devices was very unreliable, as the operation principle was unclear at that time, mainly because of lack of a basic theory. In fact, the basic physic for understanding the phenomena were not established yet, as they were still under investigation by the Physicist involved in the emerging theories of Wave or Quantum Mechanics.

---

[3] Braun demonstrated the properties of semiconductors to an audience at Leipzig on November 14, 1876, but it found no useful application until the advent of radio in the early 1900s when it was used as the signal detector in a 'crystal radio' set.

Indeed, the cat's whisker detector and other semiconductor-based devices, were unconscious implementations of the rectifying metal-semiconductor junction, whose operation principle would be fully understood only several years later, in 1938, by Walter Schottky, who explained the basic physics of the interface between metals and semiconductors, caused by the energy barrier that was formed at the interface itself.

New technology and new equipment, however, were needed to further support the growth of the radio. This ultimately led to the development of the vacuum tube technology. The first device was the diode valve discovered by John Ambrose Fleming, a Marconi's co-worker. It consisted of a heated element in an evacuated glass bulb; a second element was also placed in the bulb but not heated. It was found that an electric current only flowed in one direction with electrons leaving the heated cathode and flowing towards the second element called the anode, and not in the other direction.

In the USA Lee de Forest, replicated Fleming's diode and went a step ahead by adding an additional element to give a device he called Audion. Although de Forest applied for several patents in the years between 1905 and 1907, the invention of the triode is normally taken to be 1906. Initially the triode was only used as a detector as its operation was not fully understood, and this prevented its potential from being utilised. It took some time before the full potential of the triode was realised. Eventually, it was de Forest himself who succeeded in using it as an amplifier and in 1912 he built an amplifier using two devices. Since then, tubes with grids could be used for many purposes, including amplification, rectification, switching, oscillation, and frequency conversion.

Although thermionic tubes enabled far greater performance to be gained in radio receivers, the performance of the devices was still very poor and receivers of that time suffered from lack of sensitivity and poor selectivity.

The solution came in 1918 by Edwin Armstrong who developed a receiver where the incoming signal was converted down to a fixed intermediate frequency, where it could be amplified and appropriately filtered.

A second major contribution by Armstrong was the invention of the wideband Frequency Modulation (FM) in 1934; here, rather than varying ('modulating') the amplitude of a radio wave to encode an audio signal, the frequency was altered (modulated) proportionally to the amplitude of the audio wave itself.

In figure 4, a classic examples of the vacuum tube technology is shown; on the left, a vacuum valve, possibly a triode is represented. The anode is clearly visible, surrounding the other electrodes.

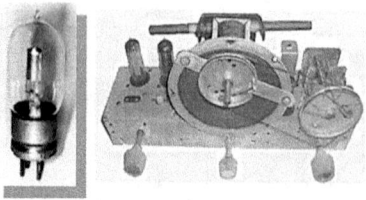

Fig. 4 – (*left*) An early example of a vacuum tube in '20s. (*right*) A radio receiver set.

On the right a classical superheterodyne radio receiver set is shown; two valves are visible, along with the variable tuning capacitor and perhaps the IF filter in the classical metallic enclosure.

Vacuum tubes remained the preferred amplifying device for 40 years, until researchers at Bell Labs invented the transistor in 1947. In all those years the development of all the techniques related to the development of what today we call electronic devices, components and circuits was mainly, if not almost exclusively, dedicated to the progress of radio communication.

The invention of Radar

Radio communications and the related technologies continued evolving in the subsequent decades, improving transmitted power, receiver sensitivity and moving to higher frequencies.

A limit to those progress of communications, however, was set by the intrinsic limitations of vacuum tube technology itself. One limitation, in particular, had great relevance as it was related to attain simultaneously transmission at frequency in the range of centimetres and peak powers of hundreds or even thousands Watt.

One of the major applications requiring such a technology was related to the possibility of detecting metallic object located at great distance with respect to the transmitter, while obtaining at the same time information both on distance and bearing, the first requiring the transmission of short pulses at high peak power and the second requiring transmission at short wavelengths, to keep antennas dimension at reasonable size.

The idea of using electromagnetic waves for object detection was suggested by some observations made by Marconi himself, who noticed radio waves were being reflected back to the transmitter by objects in his 1933 radio beacon operating at 90 centimetres conducted between Rome and Castel Gandolfo.

At that time technology was not mature to implement such a system similar to the radar that we know today, due to other limitations in addition to lack in power transmitters like, by instance, the fact that fast switching from transmit to receive mode to enable an antenna to be switched had not been developed yet.

Radio-detection[4] systems was secretly developed by several nations in the period before and during World War II. The term radar was coined in 1940 by the United States Navy as an acronym for RAdio Detection And Ranging.

Great Britain was one of the leading developers of such systems in the years leading up to World War II. The first experiments in the United Kingdom in 1935 were performed by Robert Watson-Watt[5] who demonstrated the principle by detecting a Handley Page bomber at a distance of 12 km using the BBC shortwave transmitter at Daventry.

Instead waiting for technology to become mature, which would eventually result in having it too late or possibly not having it at all had the Germans overcome, Watson Watt decided to take advantage of the technology available at that time, i.e. 30 MHz high power transmitters used for broadcasting by BBC; he overcame the inherent limitations in lateral resolution capability determined by the wavelength, implementing an original configuration with one transmitter and two bistatic receivers. Exact synchronization between the receivers led to detecting both rang as well bearing of the target, using simple trigonometric calculations, resulting in the so called bistatic configuration.

This solution envisaged by Watson Watt led to the implementation of the early warning radar[6] system made of stations around the British Isles to provide warning of air raids,

[4] The term RADAR was coined in 1940 by the United States Navy as an acronym for RAdio Detection And Ranging.

[5] Initially appointed by the Air Ministry in 1934 to develop a weapon in response to rumours of a German death ray machine, Watson-Watt instead began experimenting on the potential use of RADAR.

[6] The term radar is somewhere used in the text, for the sake of simplicity, to designate the earl radio-detection systems, even if the term was not invented yes.

called *Chain Home*, in time to be used since the early days of WWII. Bistatic radar systems gave then the way to monostatic systems. The monostatic systems were much easier to implement since they eliminated the geometric complexities introduced by the separate transmitter and receiver sites. In addition, aircraft and shipborne applications became possible as smaller components were developed.

In Italy in the '30, two scientists pioneered advancements in radar technology[7], both working in the Regia Marina Research Centre Located in Livorno. Professor and Ugo Tiberio published the radar equation, a theoretical study on the operation of radar[8]; Professor Nello Carrara investigated radio valves for transmitting and detecting microwaves[9].

In figure 5, the abstract of a paper published by Professor Carrara in 1932 on detection of microwaves[10] is shown.

Proceedings of the Institute of Radio Engineers
Volume 20, Number 10   *October, 1932*

THE DETECTION OF MICROWAVES*

BY

NELLO CARRARA

(Regio Istituto Elettrotecnico e delle Comunicazioni della Marina, Leghorn, Italy)

Summary—*The results of researches made to recognize the best conditions in which triodes can be used for the detection of microwaves (frequencies of about 10^9 per sec) are reported here. It seems that the detecting triodes, which must have their grids at a very high positive potential and the anode at a potential just lower than that of the positive end of the wire, act simply like rectifying diodes with electrodes very near to one another.*

Fig. 5 – Abstract of the paper published by Professor Carrara in 1932, where the original term microwave was coined.

Despite the brilliant success of *Chain Home* the efforts to implement a 'high frequency' or microwave high power valve did not stop. Soon after the war started in 1939, the British Admiralty issued a research contract for the development of vacuum tubes for 10 cm wavelength (3 GHz) with the Physics Department at Birmingham University for transmitting tubes, where Randall and Boot invented the multicavity magnetron, see figure 6.

This development was to revolutionise radar and was the key element in making microwave radar possible.

Fig. 6 – First Magnetron prototype developed at Birmingham in 1940.

---

[7] Unfortunately, poor understanding by senior officers and budget limitations prevented Regia Marina from having access to a strategic equipment of naval warfare.

[8] U. Tiberio, *'Misura di distanze per mezzo di onde ultracorte' rivista*, Alta Frequenza My 1939 (Stato Maggiore della Marina, Notiziario della Marina n. 10 Ottobre 1998).

[9] N. Carrara, *The detection of Microwaves*, Proceedings of IRE, vol. 20, no. 10, October 1932.

[10] The term microwave was coined by Professor Nello Carrara.

The magnetron is basically a metallic cylinder, the anode collinear with the centre of the cylinder, with a number of cavities acting as microwave resonators, the cathode and a magnet generating a magnetic field parallel to the axis of the cylinder, see figure 7.

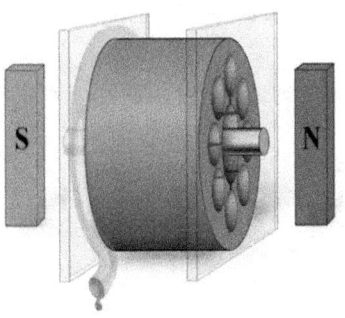

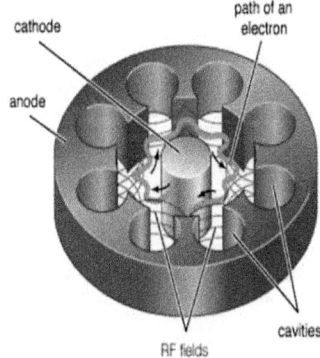

Fig. 7 – Basic structure (*top*) and operation principle (*bottom*) of Magnetron.

Under the combined action of a dc electric field and an orthogonal magnetic field, a cloud of electrons leave the cathode and are accelerated toward the anode with different paths, depending upon the initial conditions.

As this cloud, grouping in spokes, comes in proximity to the anode, it interacts with the r.f. fields at the cavities, experiencing lead-lag in velocity depending on the polarity of the field itself.

As a results of the interaction with the cavities, the spoke pattern will rotate to maintain its presence in an opposing field.

The consequent 'automatic' synchronism between the electron spoke pattern and the r.f. field polarity determines the feedback to maintain stable oscillation.

Using magnetron yielded three distinct advantages with respect to *Home Chain* solution[11] in terms of resolution, since a microwave antenna could be made highly directional.

---

[11] Despite suboptimal, the *Home Chain* solution deserves much credit as it represented the world's first integrated air defence system; the network was continually expanded, with over forty stations operational by the war's end.

The magnetron became the most diffused technology for the implementation of radar transmitter, followed by the invention of Magnetron, the Klystron was invented by Varian Brothers and, later on, the Travelling way tube (TWT).

It will take many decades before solid state amplifiers could replace, at least for some applications, those powerful precursors based on vacuum tube technology.

Alan Turing and Claude Shannon: the Computing Machine and the Digital Communications

Since the early '40s the continuing progresses in the field of computing and communications pushed ahead the development of new concepts for the implementation of electronic technologies to overcome the limitations of vacuum tube technologies.

Although that process was the result of the work of many scientists, two of them deserve a particular credit. One is Alan Turing who conceived the basic idea of the computing machine and the other is Claude Shannon who started the era of numeric communications.

Alan Turing and the Computing Machine

In 1936 Alan Turing, a Fellow of King's College, Cambridge in his paper entitled *On Computable Numbers, with an Application to the Entscheidungsproblem*, whose abstract is represented in figure 8, was able to devise a powerful yet simple model of what a computer could be. Indeed his model proved so useful and elegant that it has provided the standard definition of computability. The intent of Turing was to give an answer to David Hilbert's famous *Decision problem*[12]; the point was to establish whether it is in principle possible to find an effectively computable decision procedure which can infallibly, and in a finite time, reveal whether or not any given proposition is provable[13].

ON COMPUTABLE NUMBERS, WITH AN APPLICATION TO
THE ENTSCHEIDUNGSPROBLEM

By A. M. Turing.

[Received 28 May, 1936.—Read 12 November, 1936.]

The "computable" numbers may be described briefly as the real numbers whose expressions as a decimal are calculable by finite means. Although the subject of this paper is ostensibly the computable *numbers*, it is almost equally easy to define and investigate computable functions of an integral variable or a real or computable variable, computable predicates, and so forth. The fundamental problems involved are, however, the same in each case, and I have chosen the computable numbers for explicit treatment as involving the least cumbrous technique. I hope shortly to give an account of the relations of the computable numbers, functions, and so forth to one another. This will include a development of the theory of functions of a real variable expressed in terms of computable numbers. According to my definition, a number is computable

Fig. 8 – The original paper entitled *On computable numbers...*, published by Turing on 1936.

Having defined his notion of a computing machine, Turing showed that there exist problems, notably the famous 'Halting Problem' for Turing Machines, that cannot be effectively computed by this means.

---

[12] *Entscheidungsproblem* in German.

[13] An 'effectively computable' procedure is supposed to be one that can be performed by systematic application of clearly specified rule.

Starting from sophisticated and abstract issues of mathematical logic, Turing had devised a simple and elegant machine[14] that can simulate the logic of any computer algorithm. In other terms, anything a real computer can compute, a Turing machine can also compute.

According to Turing: «[...] it is possible to invent a single machine which can be used to compute any computable sequence. If this machine U is supplied with the tape on the beginning of which is written the string of quintuples separated by semicolons of some computing machine M, then U will compute the same sequence as M».

The Turing Machine consists of a long tape divided into squares (see fig. 9), onto which symbols can be written and later erased, together with a read/write head (as in a tape recorder) which can write (or erase) symbols and also read them. In order to 'remember what it is doing', the Turing Machine has a very limited memory in the form of a 'state', which can take any of a specified – and finite – range of values. One of these is the beginning state, from which computation starts.

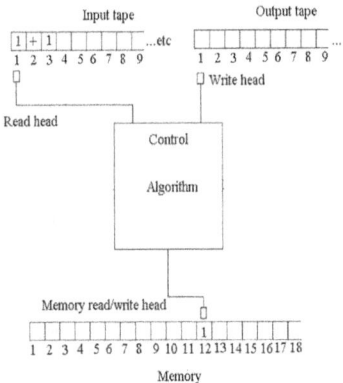

Fig. 9 – Basic operation of the Turing Machine: the addition of two numbers.

This seems an extremely limited repertoire of operations, but amazingly, it proves sufficient for calculating anything that we know how to calculate. By designing an appropriate machine table, a Turing Machine can be made to calculate anything that is calculable by even a modern computer. The revolutionary contribution by Turing is to implement on single machine that could perform every task, rather to implement a different machine for every specific task.

Turing was also a cryptanalyst. During WWII, Turing worked at Bletchley Park, Britain's Code-breaking Centre. He devised a number of techniques for breaking German ciphers, including improvements to the pre-war Polish bombe method and an electromechanical machine that could find settings for the Enigma machine. Turing played a pivotal role in cracking intercepted coded messages that enabled the Allies to defeat Germany in many crucial engagements, including the Battle of the Atlantic.

After the war, he worked at the National Physical Laboratory, where he designed the ACE[15], among the first designs for a stored-program computer. In 1948 Turing joined Max

[14] Turing Universal Machine.

[15] The Automatic Computing Engine (ACE) was an early electronic stored-program computer design produced by Alan Turing at the invitation of John R. Womersley, superintendent of the Mathematics Division of the National Physical Laboratory (NPL), in Teddington-London.

Newman's Computing Laboratory at the University of Manchester, where he helped develop the Manchester computers and became interested in mathematical biology.

Claude Shannon and the Sampling Theorem

In 1928 Harry Nyquist of Bell Telephone Laboratories ('Bell Labs') presented to the American Institute of Electrical Engineers a paper entitled *Certain topics in telegraph transmission theory*. In this paper he laid the groundwork for how samples of a time-varying signal could be used to exactly recreate the original signal, if certain conditions are met.

In 1949 Claude Shannon, also at Bell Labs, wrote a paper entitled *Communication in the presence of noise* where he presented the first formal proof of the general concept presented by Nyquist. The conditions specified by Nyquist for recovering the original signal from samples of the signal are now known as *the Nyquist-Shannon Sampling Theorem*.

Shannon's original statement of Sampling Theorem reads: «If a function contains no frequencies higher than ω cps, then it is completely determined by giving its ordinates at a series of points spaced $1/2\omega$ apart».

The principal impact of the Nyquist-Shannon sampling theorem on Information Theory is that it allows the replacement of a continuous band-limited signal, referred to as analogue systems, by a discrete sequence of its samples, referred to as sampled data systems, without the loss of any information. Also it specifies the lowest rate, or Nyquist[16] rate, that makes possible to recover the original signal. Higher rates of sampling do have advantages for establishing bounds, but would not be necessary for a general signal reconstruction.

To communicate a time-varying analogue signal using only periodic samples of the signal offers two major benefits.

First, if a communications channel supports rapid transmission of individual samples, then once a sample has been transmitted for a given signal, the communications channel can be used to transmit samples of other signals until it is time to again transmit a sample of the first signal[17].

Secondly, if a communications channel inherently transmits information in a digital manner (e.g., using bits), then converting the individual samples for a given signal to numerical values for transmission represents the most direct method for communicating the time-varying analogue signal using an inherently digital-type communications.

The two issues out-lighted before, multiplexing and numeric coding, represent the basis of modern digital communications, increasing both channel capability and accuracy, at the expenses of increased complexity and speed of the electronic circuits needed to implement the related configurations.

This has represented a formidable boost to the evolution of Electronics that, starting from the transistor and first small scale integrated circuits (ICs), ultimately evolved in the Very Large Scale of Integration (VLSI) technology.

Though more complex and costly than their analogue counterpart, the sampled data systems exhibit considerable advantages, as many of the algorithms employed are a result of developments made in the area of signal processing are in some cases capable of functions unrealizable by current analogue techniques. With increased usage, a growing demand has evolved to understand the theoretical basis required in interfacing these sampled data-systems to the analogue world.

---

[16] H.T. Nyquist discovered the Sampling Theorem while working for Bell Labs, and was highly respected by Claude Shannon.

[17] This is called 'time-division multiplexing' (TDM) and requires that the receiver be able to separate the samples appropriately so as to reconstruct the various independent signals (the telephone system uses this technique to multiplex numerous calls onto one trunk line).

From the Light Bulb to Crystals; the Dawn of Electronic Age

The quantum theory of solids was fairly well established by the mid-1930s, when semiconductors began to be of interest to scientists seeking solid-state alternatives to vacuum-tube amplifiers and electromechanical relays.

As mentioned above, the pioneering steps were performed by two German inventors. Lilienfeld who patented the original idea of field effect transistor and Oskar Heil, who patented the first FET in 1935.

The time, however, were not mature for the exploitation of those early ideas, mainly for the lack of theoretical basis and limitations of the available technology.

The Invention of the Transistor

There had been a few solid-state electronic devices until the end of '30s, the few poorly understood, until the work of Nevill Mott and Schottky in 1939 showed the phenomenon related to interfacing semiconductors were due to asymmetric potential barrier at the interface.

In late 1939 and early 1940, Shockley and Walter Brattain tried to fabricate a solid-state amplifier by using a third electrode to modulate this barrier layer, but their primitive attempts failed completely. Only later, in 1947 John Bardeen, Brattain and Shockley at Bell Labs succeeded in fabricating the working point-contact transistor (see fig. 10). This invention stimulated a huge research effort in solid state electronics. For their achievements, Bardeen, Shockley and Brattain received the Nobel Prize in Physics, in 1956.

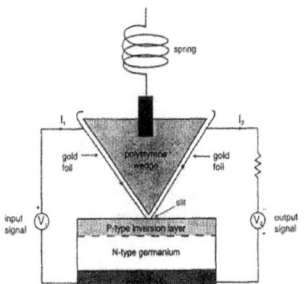 

Fig. 10 – Schematic diagram Photograph (*left*) and (*right*) of the first point- contact transistor.

After that, Shockley conceived a distinctly different type of transistor based on the p-n junction discovered by Russel Ohl in 1939, Shockley had recognized the role of minority carriers in the transport mechanism, and his analysis concluded with the invention of a the junction transistor, a sandwich of lightly doped n-type material between two regions of p-type, or the other way around. With one p-n junction forward biased and the other reverse biased, minority carriers would be injected from the forward-biased junction into the n-type material sandwiched region. They could then diffuse across the n-type region and, if the region was thin enough, a large fraction would be collected at the reverse junction.

Thus, current generated in a low impedance circuit, the emitter, would create a similar current flow in a high impedance circuit, the collector, and power gain would result

In figure 11 Shockley is portrayed with a blackboard behind him where a sketch of a bipolar junction transistor BJT is drawn, depicting the energy band diagram and Fermi energy to explain the operation principle. The development of planar technology was the subsequent and decisive step ahead. Although planar technology involves more process steps than using the alloy diffusion method, many transistors, or whole integrated circuits, can be formed at the same time the same wafer. As several wafers of silicon can be processed at the same time, the planar process resulted much more economical than the alloy diffusion method, thus opening the way to era of Integrated Circuits (ICs) era.

Fig. 11 – Shockley at blackboard illustrating the basic operation of the junction transistor.

Following the invention of JFET in 1957, in 1960 Dawon Kahng and Martin Atalla invented the MOSFET (Metal Oxide Semiconductor).

As a further and decisive development, In 1963, while working for Fairchild Semiconductor, Frank Wanlass[18] patented the Complementary Metal Oxide Semiconductor (CMOS) featuring two important characteristics, high noise immunity and low static power consumption; in fact, since one transistor of the pair is always off, the series combination draws significant power only momentarily during switching between transition from on to off states.

CMOS, also due to the scalability allowed for by the technology, represented the basis for the evolution of the technology in the following years.

The Invention of the Integrated Circuit

For many years, transistors were made as individual electronic components and were connected to other electronic components on boards to make an electronic circuit. They were much smaller than vacuum tubes and consumed much less power. However, it did not take long time before the limits of this circuit construction technique were reached. Circuits based on individual transistors became too large and too difficult to assemble. Furthermore, the transistor circuits where prone to time delays for electric signals to propagate a long distance in these large circuits. A new solution was needed to overcome.

The turning point occurred in 1958 when Jack Kilby envisaged a solution to the problem of large numbers of components, and built the first integrated circuit[19], represented in the picture of figure 12, for which in 2000 he was awarded the Nobel Prize for Physics. The

[18] In 1991, Wanlass was awarded the IEEE Solid-State Circuits Award.
[19] The circuit was a flip flop and was fabricated at Texas Instruments.

next year Robert Noyce at Fairchild developed a similar solution. The Integrated Circuit was born: instead of fabricating transistors individually, a multiplicity of transistors could be made at the same time, on the same substrate of semiconductor (wafer), including not only transistors, but other electric components such as resistors, capacitors and diodes.

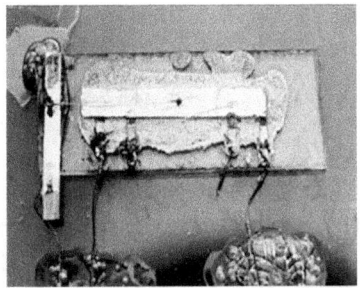

Fig. 12 – The first Integrated circuit fabricated by Kylby.

Few years later, in 1963 Frank Wanlass, while working at Fairchild Semiconductors, invented and patented the first CMOS logic circuit. The foundations were laid-down for the development of modern Electronic technology which ultimately resulted in VLSI circuits.

The VLSI Technology

With the advent of the integrated circuits, the electronic circuit development has been accomplished with the downscaling of component size. Scaling transistor's dimensions has been the key issue for the development of silicon integrated circuits. The more an IC is scaled, the higher its packing density and the lower its cost. These have been key achievements which, dramatically reduced cost per function, and much reduced physical size compared to the previous technologies.

The paradigm of scaling includes two different approaches, constant field scaling and constant voltage scaling, often mixed in practical implementations.

In constant field scaling the physical dimensions of the device and the supply and threshold voltages, are reduced by the same factor k; accordingly, the two-dimensional pattern of the electric field is maintained constant, while circuit density increases by the factor k^2. However, it requires a reduction in the power supply voltage as one decreases the minimum feature size.

Constant voltage scaling does not have this problem and is therefore the preferred scaling method since it provides voltage compatibility with older circuit technologies. The disadvantage of constant voltage scaling is that the electric field increases as the minimum feature length is reduced. These remarkable results out-light the power of scaling. For example, scaling by 5 provides 25 more circuits with the same chip size and results in 5 performance increase with no increase in chip power.

The first integrated circuits contained only a few transistors. Early digital circuits containing tens of transistors provided a few logic gates, and early linear ICs had as few as two transistors, and were termed as Small Scale of Integration (SSI) circuits.

The next step in the development of integrated circuits, occurred in the late 1960s, when devices were introduced which contained hundreds of transistors on each chip, called 'Medium-Scale Integration' (MSI).

The number of transistors in an integrated circuit has increased dramatically since then (see fig. 13) ultimately leading to 'Large Scale of Integration'. This was the final step in the development process started in the 1980s and continuing to day, which represents the first age of Integrated Circuits, while the second age started with the introduction of the first microprocessor, the 4004 by Intel in 1972 followed by the 8080 in 1974.

Multiple developments were required to achieve those goals. Manufacturers moved to smaller design rules and cleaner fabrication facilities, so that they could make chips with more transistors while maintaining adequate yield. The path of process improvements was summarized by the International Technology Roadmap for Semiconductors (ITRS). Design tools improved enough to make it practical to finish these designs in a reasonable time. The more energy-efficient CMOS replaced NMOS and PMOS, avoiding a prohibitive increase in power consumption.

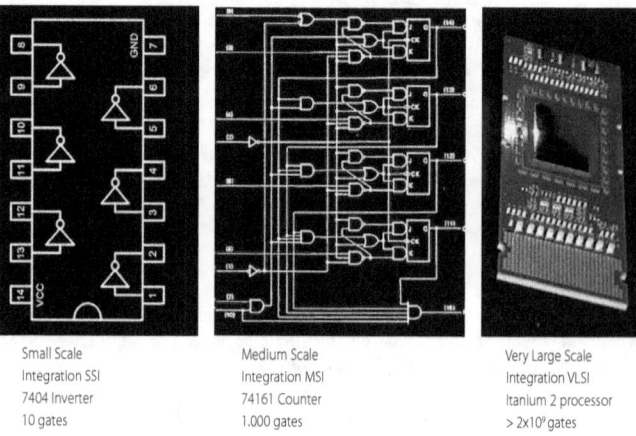

Small Scale	Medium Scale	Very Large Scale
Integration SSI	Integration MSI	Integration VLSI
7404 Inverter	74161 Counter	Itanium 2 processor
10 gates	1.000 gates	> 2x10^9 gates

Fig. 13 – Three generations of Integrated circuits, SSI, MSI and VLSI.

In 1986 the first one-megabit RAM chips were introduced, containing more than one million transistors. Microprocessor chips passed the million-transistor mark in 1989 and the several billion-transistor, nowadays. The trend continues largely unabated, with chips containing tens of billions of transistors.

The Moore's Law

In 1965 Gordon Moore, then R&D Director at Fairchild Semiconductors, in an article published on April 19, 1965, observed that the number of components in a dense integrated circuit had doubled approximately every year, and speculated that it would continue to do so for at least the next ten years[20]. The phrase 'Moore's Law' was popularised by the press and became the golden rule for the electronics industry.

The prediction has become a target for miniaturization, and has had widespread impact in many areas of technological change.

---

[20] In 1975, he revised the forecast rate to approximately every two years.

As a matter of fact, for more than 30 years, since the 1960's, the number of transistors per unit area has been doubling every 1.5 years.

In figure 14, the diagram corresponding to the increase in transistor density as predicted by Moore's Law is represented and compared with the development of computational power since the early years of 1900. to date. The computation capability is evaluated in IPS or instructions per second, an useful reference to compare CPU speed of different computer.

The Moore's Law trend in represented by red line, and is extrapolated as a function of time from early 1900 to a foreseeable future, the area in light blue grey; it turns out that nowadays, MIPS of modern computers are in the range of several hundred thousand MIPS.

For comparison Colossus , the machine which helped to break the ENIGMA code, was in the range of 1 IPS, while the Apple II was less than 100,000 IPS.

A comparison is also made with the equivalent brain capability of animals and humans again in figure 14. Various estimates place a brains processing power at 1 trillion MIPS.

Questions about the end of CMOS scaling have been discussed, but the predictions have proven wrong. The most spectacular failures in predicting the end were related to the inherent limitations of the so called 'lithography barrier', in which it was assumed that spatial resolution smaller than the wavelength used for the lithographic process was not possible, and the 'oxide scaling barrier', in which it was claimed that the gate oxide thickness cannot be reduced below 3 nm due to gate leakage. Both the predictions failed and scaling was pushed beyond the limits defined by those predictions.

Today the most powerful microprocessor represents roughly the equivalent brain power of a mouse. Extrapolating Moore's Law from here it will be somewhere around 2022 that the most powerful CPU will have the equivalent brain power of a human.

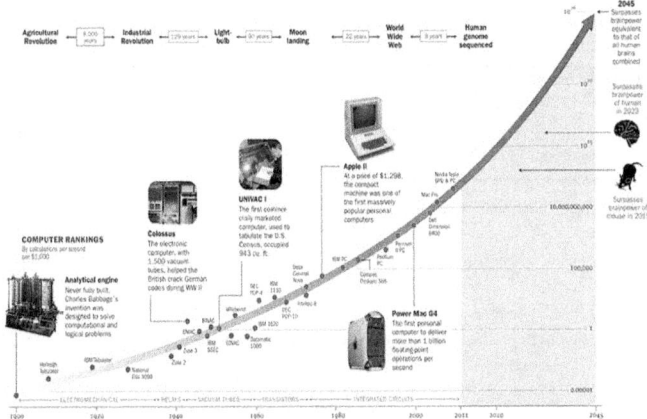

Fig. 14 – The Moore's Law as compared to the computational capacity of animal and humans in MIPS.

Integrated Circuits at Microwave and Millimetre Wave Frequencies

Designing analogue circuits to operate at radio frequencies (RF) has traditionally been an iterative and time-consuming process, in comparison with digital logic design, and remains very much the domain of a very specialised community of engineers.

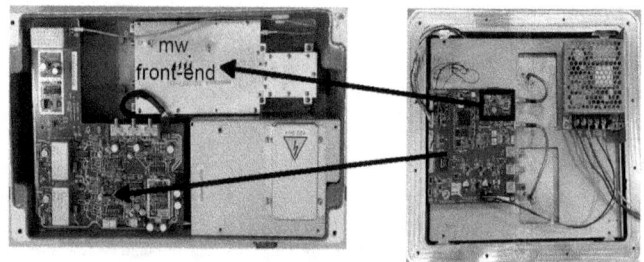

Fig. 15 – Comparison of two generation of microwave subsystem, outlighting the impact of MMIC.

The development of Monolithic Microwave Integrated Circuits (MMICs) has represented a breakthrough in microwave design and fabrication and the most important development in microwave technology during the last decades.

MMICs aim at integrating on one single chip, all the active and passive components, including planar transmission lines, needed to perform complex functionalities like, amplification, mixing, VCOs, phase detectors and in general all the major functionalities needed for implementing the modern transceivers that equip mobile phones, GPS receivers, not to mention but a few.

The acronym MMIC appeared for the first time in 1975in a paper published by Ray Pengelly and James Turne at Plessey, entitled *Monolithic Broadband GaAs F.E.T. Amplifiers*. They used computer optimization to design their lumped element.

The commercial pressures for more bandwidth grew very rapidly during the early nineties, with the massive growth in mobile telephone use, and the growth in the Internet. The possible limitations of CMOS at frequencies above 1 GHz, were considered a limit at that time, so that alternative materials were explored for fabricating MMIC, like, by instance, Gallium Arsenide (GaAS), Indium Phosphide (InP) and Gallium Nitride (GaN).

CMOS technology, despite the scepticism of the years '90, proved to be successful in the fabrication of circuits well in the microwave range. An example of the dramatic improvement allowed for by the combined MMIC and VLSI technologies is represented in fig. 15, where two generations of the same equipment, a Dedicated Short Range Communication[21] (DSRC) system for electronic toll payment operating at 5.8 GHz are represented.

On the other hand, new types of transistors have emerged for specific applications at high frequencies, in particular for fabricating power amplifiers, such as heterojunction bipolar transistors (HBTs), metal-semiconductor FETs (MESFETs), high electron mobility transistors (HEMTs), and laterally diffused MOS (LDMOS). These transistors take advantage of the materials to produce the best amplifying and power handling capability.

The Internet of Things

In its 1999 vision statement, the European Union's Information Society Technologies Program Advisory Group (ISTAG) used the term *Ambient Intelligence* (AmI) to describe a vision where «people will be surrounded by intelligent and intuitive interfaces embedded in everyday objects around us and an environment recognizing and responding to the presence of individuals in an invisible way».

---

[21] DSRC provide communications between a vehicle and the roadside in specific locations, for example toll plazas.

Ambient Intelligence represents the convergence of three key technologies: Ubiquitous Computing, Ubiquitous Communication and Intelligent User, which ultimately resulted in the vision of Internet of Things (IoT).

IoT represents the convergence of inexpensive sensors, self-powered wireless communication and embedded computation units, according to the paradigm of Wireless Sensor Networks (WSNs), a technology emerged on 2003. Using tens to thousands of those devices allows mapping physical and environmental parameters at an unprecedented time/space scale with an revolutionary potential.

This technology is already affecting all aspects of our lives and will even more in the future. Substantial improvements can be achieved in agriculture, industrial automation, transportation, energy and medicine.

An example of large scale application of WSN technology to the real-world is represented by the implementation of a distributed real-time monitoring system first deployed in eni-Versalis chemical plant located in Mantua in 2010, the followed by a even more installation in the eni refinery located in Gela, Sicily.

The system consists of tens of units deployed in critical points within the plant and along the plant's perimeter, each equipped with both VOC and H2S detectors. The units are connected to gateways forwarding data at minute rate to the plant Security Centre. Among many other functionalities, the system is capable to display the status of the data collected for analysis at a glance of the current concentration levels over the whole area of the plant (see fig. 16). The concentration measurements are correlated to wind speed and direction, represented by blue arrows, to precisely identify the source of possible leaks[22].

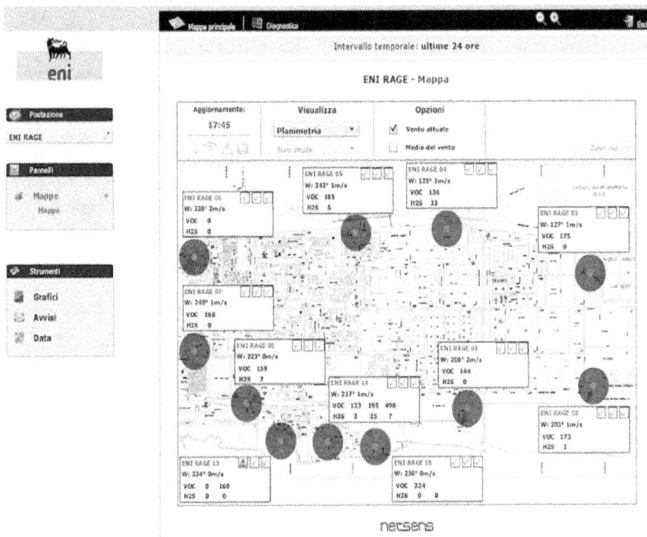

Fig. 16 – An implementation of Internet of Things: the concentration mapping of VOC and H2S in a refinery.

[22] G. Manes et al., *Real-time gas emission monitoring at hazardous sites using a distributed point-source sensing infrastructure (invited)*, Special Issue *Towards Energy-Neutral WSN Architectures: Energy Harvesting and Other Enabling Technologies*, Sensors ISSN 1424-8220, 2015.

From More Moore to More than Moore

The continued shrinking of physical feature sizes of the digital functionalities is internationally referred to as the 'More Moore' approach.

In recent years a second concept arose, referred to as 'More than Moore' (MtM) denoting a set of technologies that enable non digital micro/nano-electronic functions. They are based on, or derived from, silicon technology but do not necessarily scale with Moore's Law.

MtM devices typically impact the area or sensors and transducer; in general they provide conversion of non-digital as well as non-electronic information, such as mechanical, thermal, acoustic, chemical, optical and biomedical functions, to digital data and vice versa.

A typical example of MtM technology is given by Micro-Electro-Mechanical Systems (MEMS). MEMS, is a technology that can be defined as miniaturized mechanical and electro-mechanical devices that are made using the micro-fabrication techniques based on technologies largely imported by the semiconductor industry. Like ICs, MEMS basic techniques are deposition of material layers, patterning by photolithography and etching to produce the required shapes.

MEMS find wide application in various areas including automotive, notably in airbag systems , vehicle security. inertial brakes, automatic door locks and active suspensions, to name a few. Other important areas are seismic detection, inkjet printers and accelerometers.

An example of application of MEMS technology to very popular devices of the everyday life, according to the concept of AmI, is the gravity and accelerometer sensor used in smartphones and tablet so that images are always displayed upright, irrespectively on the orientation of the screen.

Conceptually, an accelerometer behaves as a damped mass on a spring. Under the influence of external accelerations the proof mass deflects from its neutral position. The displacement is then measured to give the acceleration. Traditionally devices would not be compatible with a modern mobile device, both in terms of size and cost.

The MEMS accelerometer represents the viable solution, currently used in modern smartphones and tablets.

The accelerometer is represented in figure 17 where the schematic operation, represented at right end, is compared to the close up image of the actual device as obtained by an electronic microscope; the circle represents in the same scale the average diameter of human hair, for reference.

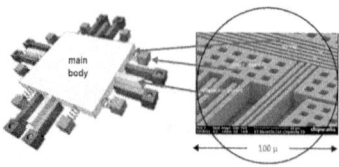

Fig. 17 – Schematic diagram (*left*) and electronic scanning image (*right*) of the MEMS accelerometer.

The accelerometer depicted in figure 17 can measure both the gravitational as well the inertial force. Measurement of gravitational mass is provided by a spring-mass system, schematically represented in the figure; the actual implementation in the MEMS domain consists of an array of tiny layer of silicon bounded to the main body. Under the effect of the gravity, the spring's geometry is altered and slightly modifies the relative position of the electrodes of a capacitor, thus allowing read-out with standard techniques. Individual gravity sensors, similar to what described above are positioned around the main body; Combining the vectorized read-out yields to detect the orientation of the sensor with respect to the direction of the gravity force.

Beyond Moore's Law

At a recent event celebrating the 50th anniversary of Moore's law, Intel's 86-year-old chairman emeritus said his law would eventually collapse, but that 'good engineering' might keep it afloat for another five to ten years'.

To take computing and communications beyond Moore's Law, however, will require entirely new scientific, engineering, and conceptual frameworks, both for computing machines and for the algorithms and software that will run on them.

So far, scaling offers so dramatic benefits that no alternative technologies can compete with. In addition, more attention and knowledge have been focused on the investigation of MOS technology. Despite the continues effort and investments, scaling CMOS is unquestionably approaching its limits. Although many issues have been resolved, scaling still cannot progress past the size of the molecule. The questions of what technology might surpass CMOS have come out. By now, there are many alternative devices that show promising replacement to CMOS for the future .

Among the many, it is worthy to mention nanodevices as carbon Nanotubes Field-Effect Transistors (CNTFETs), Nanowire Field-Effect Transistors (NWFETs), Single Electron Transistor (SET), Resonant Tunneling Diodes (RTDs), and quantum dots (QD), a nanocrystal made of semiconductor materials that is small enough to exhibit quantum mechanical properties.

To guess the candidate which would ultimately replace the powerful CMOS technology, which has dominated the scene, and it is still doing, is a very challenging and would probably require a second *Gordon Moore* to dictate a new road map.

An Apparent Paradox: the Quantum Entanglement

An extraordinary and somehow disturbing phenomenon, was originated by a letter written by Erwin Schrödinger to Einstein in which he used the word *Verschränkung*[23] to describe the correlations between two particles that interact and then separate. He shortly thereafter published a paper where he recognized the importance of the concept, and stated: «I would not call [entanglement] one but rather the characteristic trait of quantum mechanics, the one that enforces its entire departure from classical lines of thought».

Later on, however, the counterintuitive predictions of quantum mechanics were verified experimentally. where multiple particles acts as if they are linked together in a way such that the measurement of one particle's quantum state determines the possible quantum states of the other particles even though the individual particles may be spatially separated. This leads to correlations between observable physical properties of the systems. For example, it is possible to prepare two particles in a single quantum state such that when one is observed to be spin-up, the other one will always be observed to be spin-down and vice versa, this despite the fact that it is impossible to predict, according to quantum mechanics, which set of measurements will be observed.

---

[23] Translated by Schrödinger himself as entanglement.

As a result, measurements performed on one system seem to be instantaneously influencing other systems entangled with it. A classical experiment was performed by Anton Zeilinger, an Austrian physicist and pioneer in the field of quantum information and of the foundations of quantum mechanics, who devised a way[24] to take pictures using light that has not interacted with the object being photographed. The experiment is illustrated in figure 18. On the left stimulated emission of a pair of polarization-entangled photons is illustrayed. in the right the two *entangled* picture of a cat are shown[25].

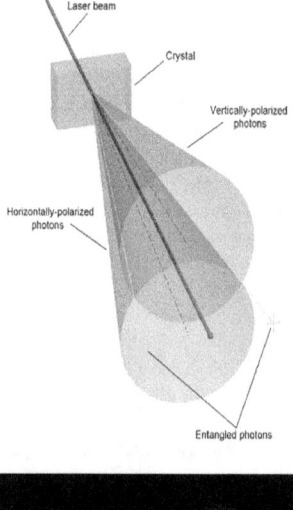

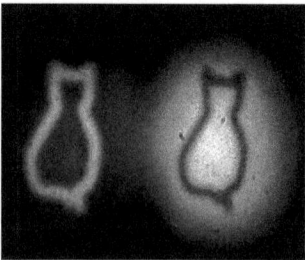

Fig. 18 – The invisible cat, achieved through the process of quantum entanglement.

Quantum entanglement is receiving much attention by the physicist and could have applications in the emerging technologies of quantum computing and quantum cryptography.

We are possibly facing a new turning point in the history of Electronics, similar to what experienced with the advent of the age of semiconductors. Only the future will say, however, whether it will be another chapter in the history of Electronics or the first chapter of a new exciting story.

---

[24] First outlined in 1991.
[25] The cat was picked in honour of a thought experiment, proposed in 1935 by the Austrian physicist Erwin Schrödinger.

Acknowledgments

The author is indebted with many Colleagues in the Department of Information Engineering of the University of Florence, for helpful discussions and comments. A particular mention deserve Professor Giuseppe Pelosi, for having unearthed the connection between the Fermi's statistics and the University of Florence, and Professors Giorgio Baccarani, Alessandro Cidronali, Stefano Selleri and Piero Tortoli for their patient and precious support in the revision of the manuscript.

The Role Played by Fermi Statistics in the Evolution of Micro- and Nanoelectronics

Giorgio Baccarani

Abstract

In this note, the role played by Fermi statistics in the evolution of Micro- and Nanoelectronics is highlighted. More specifically, the energy distribution of electrons in semiconductors is responsible for the exponential growth of the current-voltage characteristics, both in bipolar junction transistors (BJT) under weak injection conditions, as well as in field-effect transistors (FET) operating in subthreshold. The sharp transition between the off- and the on-state resulting from such an exponential behavior, made it possible to extend the scaling process from the 5 μm technology node, typical of the mid-seventies, down to the present state-of-the-art, 14-nm technology node. Throughout these four decades, the number of transistors per chip has been increasing exponentially from a few units to a few billion units, according to Moore's law.

Fermi-Dirac Statistics

In a paper[1] published first at the "Accademia dei Lincei" in February 1926 and, next, in the Zeitschrift für Physik[2] in more detailed form, Enrico Fermi derived the energy distribution of the atoms of a perfect gas enclosed within a box. Actually, he treated every atom as a three-dimensional harmonic oscillator with a characteristic frequency v, so that the quantized energy of the atoms turned out to be $(s_1 + s_2 + s_3)$ $hv = shv$, with h the Planck constant, s_1, s_2, s_3 three quantum numbers and $s = s_1 + s_2 + s_3$. However, he correctly stated that the final result would be independent of the specific assumption on the energy levels. In his derivation, he assumed that every available quantum state could be filled by just one atom, i.e. he accounted for the exclusion principle, which had just been proposed ten months earlier by Wolfgang Pauli. Indicating with $E_s = shv$ the sth energy level, he evaluated from statistical considerations the number of electrons N_s within the Q_s quantum states at the energy E_s, namely:

$$N_s = \frac{Q_s}{1 + A \exp(\beta E_s)} \tag{1}$$

with $\beta = 1/k_B T$, k_B the Boltzmann constant, and T the absolute temperature. By setting $A = \exp(-E_F/k_B T)$, eq. (1) takes the usual form

[1] E. Fermi, *Sulla quantizzazione del gas perfetto monoatomico* [*On the quantization of an idealmonoatomic gas*], «Atti dell'Accademia dei Lincei», vol. 3, 1926, pp. 145-149.

[2] E. Fermi, *Zur Quantelung des idealen einatomigen Gases*, «Zeitschrift für Physik», vol. 36, no. 11-12, 1926, pp. 902-912.

$$f = \frac{N_s}{Q_s} = \frac{1}{1 + \exp\{(E_s - E_F)/k_B T\}} \tag{2}$$

where E_F, referred to as Fermi level, represents the energy at which $f = 0.5$. In view of the exclusion principle, f can be interpreted as the occupation probability of a state at the energy E_s under equilibrium conditions. The same expression was found independently by Paul A.M. Dirac[3] in 1926, so that eq. (2) is usually called Fermi-Dirac distribution. When $E_s \gg E_F$, the asymptotic limit of (2) is $f \to \exp[-(E_s - E_F)/k_B T]$, *i.e.*, Boltzmann's distribution, which holds for classical particles not subject to the exclusion principle.

Eq. (2) is much more general than one could expect from the derivation procedure devised by Fermi: it holds for all particles with semi-integer spin, including electrons, and is therefore of major importance for the comprehension of the transport properties of both semiconductors and metallic conductors. As is well known, Microelectronics is based on silicon technology, and the Fermi-Dirac distribution plays a key role in defining nearly all device properties, as will be shown in the subsequent sections of this article.

Conduction Properties of Semiconductors

In crystalline materials, electrons are subject to an internal periodic potential. The resulting electronic states are clustered in nearly-continuous energy bands separated by band gaps. Remarkably, an energy band completely filled by electrons does not contribute to current transport, due to the symmetry of the momentum vectors, which cancel two by two. Instead, only a partially-filled band contributes to transport under the action of an external field, as the momentum distribution is skewed by the field itself. If the Fermi level falls within an energy band, most states below the Fermi level are filled by electrons, and most states above the Fermi level are empty. Therefore, electrons around the Fermi level are free to move under the influence of an external field, and the material behaves as a *conductor*. This is typically the case for metals.

In insulators and semiconductors, instead, the Fermi energy falls within a band gap. The band just above the band gap is called *conduction band*, and the band just below the band gap is called *valence band*. A remarkable property of the valence band is that its empty states behave as positively-charged particles under the influence of an external field, and exhibit particle properties, fulfilling the Newton law in the classical limit of Quantum Mechanics. These particles are called *holes*.

It should be noticed that $k_B T = 25.9$ meV at room temperature ($T = 300$ K) while a typical semiconductor band gap is of the order of one eV and a typical insulator band gap is of the order of several eV. The transition of the occupation probability f from 1 to 0 occurs in a few $k_B T$ and becomes steeper as temperature decreases. Due to the wide band gap of insulators, their conduction band is empty, and their valence band is completely filled. Therefore, no current flows across an insulator, even in the presence of large external fields. For a pure semiconductor, instead, the conduction band is nearly, but not entirely, empty, and its valence band is nearly, but not entirely, filled by electrons. Therefore, semiconductors are weakly conduct-ing materials at room temperature. At zero temperature, however, the transition of the Fermi function is abrupt, and the semiconductor behaves as an insulator.

In pure semiconductors, the Fermi level lies at mid-gap, so that the electron and hole

[3] P.A.M. Dirac, *On the theory of quantum mechanics*, «Proceedings of the Royal Society of London», vol. A112, 1926, pp. 281-305.

concentrations are rather small. For silicon, the so-called intrinsic concentration of electrons and holes is of the order of 10^{10} cm^{-3} at room temperature, i.e. many orders of magnitude below the number of atoms per cubic centimeter, which is about 5×10^{22} cm^{-3}. This negligible carrier concentration makes the conductivity of pure silicon very small. However, such properties can be deeply altered if the semiconductor is suitably doped by group III or group V elements. For instance, if a silicon atom is replaced by a boron atom, which contains only 3, rather than 4, valence electrons, the additional empty state can be easily filled by an electron, so that the boron atom becomes negatively charged, and a hole is generated.

Likewise, if a silicon atom is replaced by a phosphorus atom, which contains 5, rather than 4, valence electrons, the additional electron can be easily stripped from the phosphorus atom, which becomes positively ionized, and a conduction electron is generated. Dopant atoms can be easily inserted into the silicon crystal by ion implantation, followed by a suitable annealing procedure, by which the implanted species diffuse into silicon sites, thus becoming electrically active.

The perturbation of the periodic potential due to the presence of an impurity generates a localized state within the band gap. It turns out that acceptor states generated by group III elements have an energy just above the valence-band edge, while donor states generated by group V elements have an energy just below the conduction-band edge. Therefore, nearly all impurities are ionized according to Fermi statistics for moderate doping levels, typically below 10^{19} cm^{-3}, and an equal number of holes in the valence band, or electrons in conduction band, are generated and made available for current transport.

On the other hand, if the impurity concentration exceeds the previous limit, the position of the Fermi level within the band gap is shifted either below the valence-band edge, or above the conduction-band edge. Therefore, impurities become only partially ionized. At the same time, an impurity band is formed which merges with the valence/conduction band, leading to a bandgap narrowing and to a degenerately-doped semiconductor.

Transistor Concept

Transistors are semiconductor devices where the current flowing between two contacted carrier reservoirs is controlled by the voltage applied to a third electrode. For the case of field-effect transistors (FETs), the two reservoirs are called *source* and *drain*, and the control electrode is referred to as *gate*. The physical mechanism controlling current flow is an energy barrier interposed between source and drain. The height of that barrier depends on the gate voltage. When the latter is low, the barrier is tall, and only a small fraction of high-energy electrons within the tail of the Fermi function can overcome the barrier and flow to the drain. As the gate voltage increases, the barrier height is lowered, and a progressively larger number of electrons overcomes the barrier, thereby raising the drain current.

FETs are used as switches in digital circuits implementing the fundamental logic functions, such as NAND, NOR and NOT. Therefore, the FET basic requirements are: (*i*) a fast switching speed, (*ii*) a low leakage current, and, (*iii*) a steep transition from the off- to the on-state as the gate voltage rises. The last property is especially important to allow for voltage scaling. Due to the high-energy tail of the Fermi function, the drain current increases exponentially vs. the gate voltage in subthreshold, leading to a device characteristics of the type

$$I_D = I_{D0} \exp\left\{\frac{q(V_{GS} - V_T)}{\delta_{id} k_B T}\right\}\left\{1 - \exp\left(-\frac{q V_{DS}}{k_B T}\right)\right\} \tag{3}$$

where V_{GS} is the gate-source voltage, V_{DS} is the drain-source voltage, V_T is the threshold voltage, $I_{D0} = (W/L)\, C_{ox} \mu_n (k_B T/q)^2$ is a constant for an assigned device geometry and operating temperature, and δ_{id} is an ideality factor greater than 1. Also, W and L are the device width and gate length, respectively, C_{ox} is the oxide capacitance per unit area, μ_n is the carrier mobility and q is the electron charge.

Eq. (3) holds only for $V_{GS} < V_T$. Instead, above threshold, the drain current is approximately linear with the gate voltage. When $V_{DS} > 4k_B T/q \approx 100$ mV, the last factor in braces is about equal to 1. Hence, the off-state current $I_{OFF} = I_{D0} \exp(-qV_T/\delta_{-id} k_B T)$. By an appropriate selection of the threshold voltage, the off-state current can thus be changed by orders of magnitude, according to the specific application of the digital circuit.

On the other hand, lowering the off-state current by increasing the threshold voltage implies also a lowering of the on-state current I_{ON} for an assigned supply voltage, with a negative impact on the switching speed. The International Technology Roadmap for Semiconductors[4] (ITRS) identifies three main transistor types: high-performance (HP), with I_{OFF} = 100 nA/μm, low-operating power (LOP), with I_{OFF} = 5 nA/μm, and low-standby power (LSTP), with I_{OFF} = 10 pA/μm. The most important parameters qualifying technology performance are the intrinsic delay τ_i of a self-loaded inverter, and the energy consumption for logic operation E_d. The former parameter is defined as

$$\tau_i = \frac{C_L V_{DD}}{I_{ON}} \approx \frac{C_L V_{DD}}{g_m (V_{DD} - V_T)} \tag{4}$$

with $g_m = (\partial I_D/\partial V_{GS}) \approx WC_{ox} v_{inj}$ the device transconductance at $V_{DS} = V_{DD}$, C_L the inverter load capacitance, and V_{DD} the supply voltage. In the previous expression, v_{inj} is the injection velocity of electrons at the top of the barrier.

The energy dissipation per clock cycle is instead the sum of dynamic and static energies, i.e.

$$E_d = \alpha C_L V_{DD}^2 + I_{OFF} V_{DD} T_c = C_L V_{DD}^2 \left(\alpha + \frac{I_{OFF}}{I_{ON}} \frac{T_c}{\tau_i} \right) \tag{5}$$

where T_c is the clock cycle time and α is the activity factor, *i.e.*, the ratio between the number of complete switching operations and the number of clock cycles required for the execution of an assigned algorithm. In practice, $\alpha = 1$ only for the circuits of clock generation and distribution, and $\alpha = 0.5$ if a circuit undergoes a transition $0 \to 1$ or $1 \to 0$ at every clock cycle. For most circuits within a microprocessor, such as cache memories, $\alpha \ll 1$, as most memory cells are seldom accessed. The ratio I_{OFF}/I_{ON} is typically smaller than 10^{-4}, while the ratio T_c/τ_i is typically of the order of 15 for an inverter with fan-out = 4.

Eq. (5) shows that the total energy dissipated per clock cycle may be dominated either by the dynamic term or by the static term, depending on the activity factor. In addition, it shows that the energy dissipated per clock cycle by a logic gate is proportional to the product $C_L(V_{DD})^2$. This means that scaling down the load capacitance and the supply voltage would provide huge benefits to the overall power consumption of the microprocessor unit.

The load capacitance is reduced by miniaturizing the physical dimensions of both devices and metal interconnects. The hurdles to be overcome to scale down the supply voltage are essentially due to the need to preserve the ratio $V_{DD}/(V_{DD} - V_T)$ in eq. (4). However, if the threshold voltage is scaled down in direct proportion with the supply voltage, an exponential increase of the off-state current would result, according to eq. (3), with an increase of static power, and degradation of the I_{OFF}/I_{ON} ratio.

The ideality factor δ_{id} in eq. (3) cannot be made smaller than 1 within the stan-

---

[4] http://www.itrs.net/reports.html (2012 Update).

dard transistor concept, due to Fermi statistics. The steepest sub-threshold swing is thus $SS = [\partial \log(I_D)/\partial V_{GS}]^{-1} = 60$ mV/dec of current. Hence, the transition between 10 pA/μm and 10 μA/μm, which represents a typical value for the threshold current, requires a gate voltage change $\Delta V_{GS} > 360\text{-}400$ mV. If his condition is violated, the transfer characteristics of logic gates are degraded, and so are noise-immunity margins.

The historical benefits of device scaling have been: (*i*) increased device performance; (*ii*) reduced energy consumption per logic operation, and, (*iii*) reduced cost per function. It is thus not surprising that the semiconductor industry has devoted huge efforts to pursue device miniaturization by the development of ever more sophisticated technology processes. Physical limits did not prevent such an evolution up to now. However, it must be conceded that, after the 90 nm node, the gate-oxide thickness could not be scaled in proportion with the device lateral dimensions, as required by the scaling theory, due to gate leakage. The introduction of a novel gate stack at the 45 nm node, comprising high-κ dielectric and metal gate, relieved the leakage problem, but scaling the equivalent oxide thickness (EOT) remains a serious problem. In order to keep the required performance trends, mobility had to be boosted by the use of new channel materials (SiGe for p-type FETs) and by the application of mechanical stress to transistors (tensile stress for n-type and compressive stress for p-type FETs).

Scaling Trends

Figure 1 is an ITRS plot[4], showing the evolution of the minimum-feature sizes of silicon technology over the years. Data following year 2013 are just predicted. The upper line represents the contacted half pitch dimension of the first-level metal; the intermediate line represents the printed gate length, and the lower line represents the physical gate length. In the mid-seventies and up to mid-nineties, i.e. outside the represented time scale of this plot, the supply voltage was equal to 5V and device scaling occurred at constant voltage from the 5 μm down to the 0.5 μm nodes. Therefore, the average electric field within the channel increased by an order of magnitude, from 10^4 V/cm up to 10^5 V/cm, leading to hot-electron effects and charge injection into the gate oxide. Additional drawbacks of constant-voltage scaling are an increased power consumption per unit area and a growing current density in metal interconnects. In the subsequent decade, i.e. from the 0.5 μm node down to the 90 nm node, the supply voltage was scaled more or less in direct proportion with the device physical dimensions, from 5V down to 1.1V, according to Dennard's scaling theory[5], which ensures the invariance of power consumption per unit area, as well as a linear performance increase with the scaling factor.

After the 90 nm node, however, voltage scaling became progressively more difficult, due to the need to preserve the off-state current per unit width. Therefore, the supply voltage only scaled in the last decade from 1.1 down to 0.8V, while the device physical size was reduced from 90 nm down to 14 nm, which is to date the state-of-the-art technology. In order to set an upper limit to power consumption, it was necessary to stop the increase of the clock frequency, and system performance was improved by novel architectural features and by multiple cores on the microprocessor unit (MPU).

[5] R.H. Dennard, F.H. Gaensslen, H.N. Yu, V.L. Rideout, E. Bassous, A.R. LeBlanc, *Design of ion-implanted MOSFETs with very small physical dimensions*, «IEEE J Solid-State Circuits», vol. 9, no. 5, 1974, pp. 256-268.

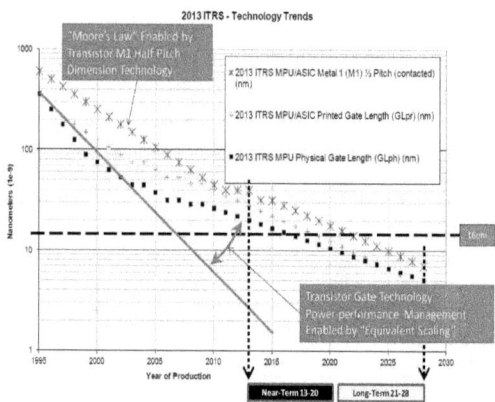

Figure ORTC2   2011 ITRS—MPU/ASIC Half Pitch and Gate Length Trends

Fig. 1 – ITRS plot[4] showing past trends and future predictions of the minimum feature sizes of semiconductor technology from year 1995 up to 2028. Asterisks: Ml contacted half pitch; dots: printed gate length; squares: physical gate lengths.

Figure 2 shows the evolution over the years of several performance indicators[6]. The upper line of symbols (upward-pointing triangles) represents the number of transistors per chip, which increases exponentially from 1 at mid-seventies up to a few billions in 2015, according to Moore's law[7]. The filled squares represent the evolution of the clock frequency: after a sustained exponential growth from 1985 to 2005, such a growth was suddenly stopped at around 3-4 GHz, due to the need to keep under control power dissipation per unit area (downward-pointing triangles) which cannot exceed 100 W/cm^2 due to the heat extraction problem. Microprocessor performance per thread, represented by filled dots, exhibits a sustained exponential growth across two decades, and still a growth, albeit at a slower pace, in the last ten years. This performance improvement per thread at constant clock frequency is mainly due to architectural improvements, such as superscalar processing, increased pipeline depth and average number of executed instructions per cycle, and increased memory-processor communication bandwidth. Finally filled diamonds represent the number of logical cores within a microprocessor unit, which has grown in the last decade from 1 to 72.

At system level, the evolution of Micro- and Nanoelectronics made it possible en enormous performance improvement of the capabilities of personal computers and smart phones. Combined with the opportunities made available by the discovery of the worldwide web, such an evolution has prompted personal communications to an extent never experienced before. Besides, the development of low-cost systems-on-chip has generated new opportunities for consumer, automotive, and industrial applications. The ultimate development expected to date is the Internet of Things (IoT), according to which the dissemination of smart sensors, assisted by distributed intelligence and suitably connected by wireless networks, is expected to provide new tools for environment and health monitoring, for control and supervision of house appliances, and for personal security.

---

[6] https://www.karlrupp.net/2015/06/40-years-of-microprocessor-trend-data/ (11/15).

[7] G.E. Moore, *Cramming more components onto integrated circuits*, «Electronics», vol. 38, no. 8, April 1965, pp. 82-85.

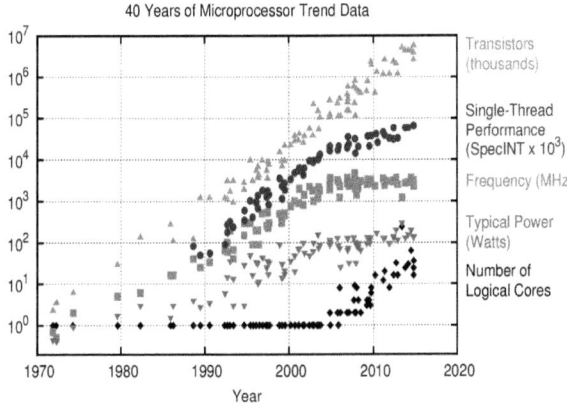

Fig. 2 – Experimental trends of several microprocessor-unit (MPU) performance indicators[6] from the early seventies up to year 2015. Upward pointing triangles: number of transistors per MPU; filled dots: single-tread performance; filled squares: clock cycle frequency; downward pointing triangles: power consumption; filled diamonds: number of logical cores.

Conclusions

The energy distribution of electrons according to Fermi statistics has been one of the key factors allowing for a controlled change of the carrier conductivity in silicon by ion implantation of suitable species pertaining to the III and V element groups. This made it possible to fabricate p-n junctions and bipolar transistors. The subsequent discovery of silicon thermal oxidation paved the way to the development of insulated-gate field-effect transistors, which turned out to be more suitable than bipolar transistors for digital applications, and are still today the workhorse of digital technology.

The evolution of Microelectronics over the last 40 years occurred primarily via device scaling. In the last decade, however, device scaling by itself provided diminished returns, and performance improvement was boosted by the use of novel materials, such as SiGe for the channel of p-type FETs, and high-κ dielectrics for gate insulators. To date, the most important requirement is no longer an improved device performance but, rather, low-power consumption. This goal is dictated by the heat-extraction problem in server applications, and by the need to preserve the battery charge in portable devices. At technology level, low power consumption can be pursued by lowering the supply voltage, which, in turn, is made possible by a steep transition from the off- to the on-state. The steepness of this transition is again ensured by Fermi statistics, the nature of which lets the drain current increase exponentially with the gate voltage at the rate of 60 mV/dec.

At present, the internal supply voltage in advanced microprocessors is 0.8V. In order to reduce it further, while preserving the switching speed, an even steeper subthreshold swing SS is required. This need had prompted several studies on novel device concepts, such as the tunnel FET (TFET), where high-energy electrons are filtered out by the bandgap in the source, or the Ferroelectric FET, where a positive feedback is introduced by a negative gate capacitance. While these novel concepts could, in principle, achieve a steeper SS, several limitations and drawbacks still prevent their industrial development.

Part II
On the Quantization of an Ideal
Monoatomic Gas
Reproduction of the original papers,
in Italian and German, by Enrico Fermi

E. Fermi

Sulla quantizzazione del gas perfetto monoatomico

from «Atti dell'Accademia dei Lincei»,
vol. 3, no. 3, 1926, pp. 145-149

ATTI

DELLA

REALE ACCADEMIA NAZIONALE

DEI LINCEI

ANNO CCCXXIII

1926

SERIE SESTA

RENDICONTI

Classe di Scienze fisiche, matematiche e naturali.

VOLUME III.

ROMA

DOTT. GIOVANNI BARDI

TIPOGRAFO DELLA R. ACCADEMIA NAZIONALE DEI LINCEI

1926

Fisica. — *Sulla quantizzazione del gas perfetto monoatomico.* Nota di ENRICO FERMI, presentata dal Socio GARBASSO.

1. Nella termodinamica classica si prende come calore specifico a volume costante di un gas perfetto monoatomico (riterendosi a una sola molecola) $c = 3 k/2$. È chiaro però che se si vuole, anche per un gas ideale, ammettere la validità del principio di Nernst, bisogna ritenere che la precedente espressione di c sia soltanto una approssimazione per temperature elevate, e che in realtà c tenda a zero per $T = 0$, in modo che si possa estendere fino allo zero assoluto l'integrale esprimente il valore dell'entropia senza l'indeterminazione della costante. E per rendersi conto del come possa avvenire una tale variazione di c, è necessario ammettere che anche i moti del gas perfetto debbano essere quantizzati. Si capisce poi come una tale quantizzazione, oltre che sul contenuto di energia del gas, avrà anche una influenza sopra la sua equazione di stato, dando così origine ai così detti fenomeni di degenerazione del gas perfetto per basse temperature.

Lo scopo di questo lavoro è di esporre un metodo per effettuare la quantizzazione del gas perfetto che, a noi pare, sia il più possibile indipendente da ipotesi non giustificate sopra il comportamento statistico delle molecole del gas.

Recentemente sono stati fatti numerosi tentativi di arrivare a stabilire l'equazione di stato del gas perfetto [1]. Le formule date dai vari autori e la nostra, differiscono tra di loro, e dalla equazione di stato classica, soltanto per temperature molto basse e per densità assai elevate; disgraziatamente sono queste le stesse circostanze nelle quali sono maggiormente importanti anche le deviazioni delle leggi dei gas reali da quelle dei gas perfetti; e siccome, in condizioni comodamente realizzabili sperimentalmente, le deviazioni dalla equazione di stato $p V = k T$ dovute alla degenerazione del gas, pur non essendo affatto trascurabili, sono sempre alquanto più piccole di quelle dovute all'essere il gas reale e non perfetto, le prime sono state fino ad ora mascherate da queste ultime; pur non essendo affatto escluso che, con una conoscenza più approfondita delle forze che agiscono tra le molecole di un gas reale, si possa, in un avvenire più o meno prossimo, separare tra di loro le due deviazioni, arrivando così a decidere sperimentalmente tra le diverse teorie della degenerazione dei gas perfetti.

(1) Vedi p. e. A. EINSTEIN, «Sitzber. d. Pr. Akad. d. Wiss.», 22 (1924), p. 261; 23 (1925), pp. 3, 18. M. PLANCK, «Sitzber. d. Pr. Akad. d. Wiss.», 23 (1925), p. 49.

2. Per poter effettuare la quantizzazione del moto delle molecole di un gas perfetto è necessario mettersi in condizioni tali da poter applicare al loro movimento le regole di Sommerfeld; e questo può naturalmente farsi in infiniti modi che, del resto, conducono tutti allo stesso risultato. Si può p. e. supporre il gas racchiuso in un recipiente parallelepipedo a pareti elastiche, quantizzando il moto, triplamente periodico, della molecola che rimbalza tra le sei faccie del recipiente; o, più generalmente, si possono assoggettare le molecole a un qualsiasi sistema di forze tale che il loro moto venga ad essere periodico e possa quindi essere quantizzato. L'ipotesi che il gas sia perfetto ci autorizza in tutti questi casi a trascurare le forze agenti tra le molecole, per modo che il moto meccanico di ciascuna di esse viene a svolgersi come se le altre non esistessero. Si può tuttavia riconoscere che la semplice quantizzazione, con le regole di Sommerfeld, del moto delle molecole, considerate come completamente indipendenti le une dalle altre, non è sufficiente per ottenere dei risultati corretti; in quanto che, pur trovandosi così un calore specifico che tende a zero per $T = 0$, si ha però che il suo valore, oltre che dalla temperatura e dalla densità, viene anche a dipendere dalla quantità totale del gas, e tende, per qualunque temperatura, al limite $3 k/2$ quando, pur restando costante la densità, la quantità totale del gas tende all'infinito. Appare dunque necessario ammettere che occorra qualche complemento alle regole di Sommerfeld, per il calcolo di sistemi che, come il nostro, contengono degli elementi non distinguibili tra di loro [1].

Per avere un suggerimento circa l'ipotesi più plausibile da farsi, conviene che esaminiamo come vanno le cose in altri sistemi che, al pari del nostro gas perfetto, contengono degli elementi indistinguibili; e precisamente vogliamo esaminare il comportamento degli atomi più pesanti dell'idrogeno, i quali tutti contengono più di un elettrone. Se consideriamo le parti profonde di un atomo pesante, siamo in condizioni tali che le forze agenti tra gli elettroni sono molto piccole in confronto di quelle esercitate dal nucleo. In queste circostanze la applicazione pura e semplice delle regole di Sommerfeld condurrebbe a prevedere che, nello stato normale dell'atomo, un numero considerevole di elettroni dovesse trovarsi in una orbita di quanto totale 1. In realtà si osserva invece che l'anello K è già saturato quando contiene due elettroni, e parimenti l'anello L si satura quando contiene 8 elettroni, etc.... Questo fatto è stato interpretato da Stoner [2], e in modo ancora più preciso da Pauli [3], al modo seguente: caratterizziamo una orbita elettronica possibile in un atomo complesso mediante 4 numeri quantici; n, k, j, m, che hanno risp. i significati di quanto totale, quanto azimutale, quanto interno e quanto magnetico. Date le diseguaglianze alle quali deb-

(1) E. FERMI, «N. C.» *1* (1924), p. 145.
(2) E. C. STONER, «Phil. Mag.», *48* (1924), p. 719.
(3) W. PAULI, «Zs. f. Phys.», *31* (1925), p. 765.

bono soddisfare questi 4 numeri, si trova che, per $n =$ 1, esistono solo due
terne di valori di k, j, m; per $n = 2$, ne esistono 8, etc.... Per rendersi
conto del fatto precedentemente osservato, basta dunque ammettere che nel-
l'atomo non possano esistere due elettroni le cui orbite siano caratterizzate
dagli stessi numeri quantici; bisogna in altre parole ammettere che una
orbita elettronica sia già « occupata » quando contiene un solo elettrone.

3. Ci proponiamo ora di ricercare se una ipotesi simile non possa dare
dei buoni risultati anche nel problema della quantizzazione del gas perfetto:
ammetteremo dunque che nel nostro gas ci possa essere al massimo una
molecola il cui movimento sia caratterizzato da certi numeri quantici, e
faremo vedere che questa ipotesi conduce a una teoria perfettamente con-
seguente della quantizzazione del gas perfetto, e che in particolare rende
ragione della prevista diminuzione del calore specifico per basse temperature, e conduce al valore esatto per la costante dell'entropia del gas perfetto.

Riservandoci di pubblicare, in una prossima occasione, i dettagli ma-
tematici della presente teoria, ci limitiamo in questa nota ad esporre i prin-
cipi del metodo seguito, e i risultati.

Dobbiamo anzitutto mettere il nostro gas in condizioni tali che il moto
delle sue molecole sia quantizzabile. Come si è visto questo può farsi in
infiniti modi; siccome però il risultato è indipendente dal modo particolare
che si adotta, noi sceglieremo quello che è più comodo per il calcolo; e
precisamente ammetteremo che sulle nostre molecole agisca una attrazione
verso un punto fisso O, di intensità proporzionale alla distanza r della molecola da O; per modo che ogni molecola verrà a costituire un oscillatore
armonico spaziale, di cui indichiamo con v la frequenza. L'orbita della molecola sarà caratterizzata dai suoi tre numeri quantici s_1, s_2, s_3, che sono
legati alla sua energia per mezzo della relazione

$$(1) \qquad w = hv(s_1 + s_2 + s_3) = hvs .$$

L'energia di una molecola può dunque prendere tutti i valori multipli
interi di hv, ed il valore shv può essere preso in $Q_s = \frac{1}{2}(s + 1)(s + 2)$
modi.

L'energia zero può dunque realizzarsi in un modo solo, l'energia hv
in 3 modi, l'energia $2hv$ in 6 modi, etc.... Per rendersi conto della influenza della ipotesi da noi fatta, che a determinati numeri quantici non
possa corrispondere più di una molecola, consideriamo il caso estremo
di avere N molecole allo zero assoluto. A questa temperatura il gas deve
trovarsi nello stato di energia minima. Se dunque non ci fosse nessuna
limitazione al numero delle molecole che possono avere una certa energia,
tutte le molecole si troverebbero nello stato di energia zero, e tutti e tre
i numeri quantici di ciascuna di esse sarebbero nulli. Invece, per la nostra

ipotesi, non ci può essere più di una molecola con tutti e tre i numeri
quantici nulli; se è quindi $N = 1$, l'unica molecola occuperà il posto di
energia zero; se è invece $N = 4$, una delle molecole occuperà il posto di
energia zero, e le altre tre i tre posti di energia $h\nu$; se è $N = 10$, una
delle molecole occuperà il posto di energia zero, altre tre i tre posti di
energia $h\nu$, e le sei rimanenti i sei posti di energia $2 h\nu$, etc....

Supponiamo ora di dover distribuire tra le nostre N molecole l'energia
complessiva $W = E h\nu$ ($E =$ numero intero); e indichiamo con $N_s \leqslant Q_s$
il numero delle molecole di energia $s h\nu$. Si trova facilmente che i valori
più probabili delle N_s sono

$$(2) \qquad N_s = \alpha \, Q_s / (e^{\beta s} + \alpha)$$

dove α e β sono delle costanti dipendenti da W e da N. Per trovare la
relazione tra queste costanti e la temperatura, osserviamo che, per effetto
della attrazione verso O, la densità del nostro gas sarà una funzione di r,
che deve tendere a zero per $r = \infty$. Per conseguenza, per $r = \infty$ debbono
cessare i fenomeni di degenerazione, e in particolare la distribuzione delle
velocità, facilmente deducibile da (2), deve trasformarsi nella legge di
Maxwell. Si trova così che deve essere

$$(3) \qquad \beta = h\nu / k T.$$

Siamo ora in grado di dedurre da (2) la funzione $n\,(\mathrm{L})\,d\,\mathrm{L}$, che ci
rappresenta, per un dato valore di r, la densità delle molecole di energia
compresa tra L ed $L + d\,L$ (Analogo della legge di Maxwell), e da questa
possiamo dedurre l'energia cinetica media $\overline{L}$ delle molecole a distanza r,
la quale è funzione, oltre che della temperatura, anche della densità n. Si
trova precisamente

$$(4) \qquad \overline{L} = \frac{3}{4} \frac{h^2 \, n^{2/3}}{\pi \, m} \, \mathrm{P} \left(\frac{2 \, \pi \, m \, k \, T}{h^2 \, n^{2/3}} \right).$$

Dove con $\mathrm{P}\,(x)$, si è indicata una funzione, di definizione analitica un
po' complicata, che, secondo che x è molto grande o molto piccolo, si può
calcolare con le formule asintotiche

$$\mathrm{P}\,(x) = x \left(1 + 2^{-\frac{5}{2}} x^{-\frac{3}{2}} + \cdots \right) \quad ; \quad \mathrm{P}\,(x) = \frac{1}{5} \sqrt[3]{\frac{9\,\pi}{2}} \left\{ 1 + \frac{5}{9} \sqrt[3]{\frac{4\,\pi^4}{3}} \, x^2 + \cdots \right.$$

Per dedurre da (4) l'equazione di stato applichiamo la relazione del
viriale. Si trova allora che la pressione è data da

$$(6) \qquad p = \frac{3}{2} \, n \, \overline{L} = \frac{h^2 \, n^{5/3}}{2 \, \pi \, m} \, \mathrm{P} \left(\frac{2 \, \pi \, m \, k \, T}{h^2 \, n^{2/3}} \right).$$

Al limite per temperature elevate, cioè per piccola degenerazione, l'equazione di stato prende dunque la forma

$$(7) \qquad p = n \, k \, T \left\{ 1 + \frac{1}{16} \frac{h^3 \, n}{(\pi \, m \, k \, T)^{3/2}} + \cdots \right\}.$$

La pressione risulta dunque maggiore di quella prevista dalla equazione di stato classica. Per un gas perfetto del peso atomico dell'elio, alla temperatura di 5° assoluti, e alla pressione di 10 atmosfere la differenza sarebbe del 15 %.

Da (4) e (5) si può anche dedurre l'espressione del calore specifico per basse temperature. Si trova

$$(8) \qquad c_v = \sqrt[3]{\frac{16 \pi^8}{9} \frac{m \, k^2}{h^2 \, n^{2/3}}} \; T + \cdots$$

Parimenti possiamo trovare il valore assoluto dell'entropia. Effettuando i calcoli si trova, per alte temperature,

$$(9) \quad S = n \int_0^T \frac{1}{T} \, d\overline{L} = n \left\{ \frac{5}{2} \log T - \log p + \log \frac{(2 \pi m)^{3/2} k^{5/2} e^{5/2}}{h^3} \right\}$$

che coincide col valore dell'entropia dato da Tetrode e da Stern.

E. Fermi

Zur Quantelung des idealen einatomigen Gases,

from «Zeitschrift für Physik»,
vol. 36, no. 11-12, 1926, pp. 902-912

ZEITSCHRIFT FÜR
PHYSIK

HERAUSGEGEBEN UNTER MITWIRKUNG
DER
DEUTSCHEN PHYSIKALISCHEN GESELLSCHAFT

VON

KARL SCHEEL

SECHSUNDDREISSIGSTER BAND
FÜNFTES HEFT
MIT 14 TEXTFIGUREN
(ABGESCHLOSSEN AM 27. MÄRZ 1926)

VERLAG VON JULIUS SPRINGER, BERLIN
1926

902

Zur Quantelung des idealen einatomigen Gases [1]).

Von E. Fermi in Florenz.

(Eingegangen am 24. März 1926.)

Wenn der Nernstsche Wärmesatz auch für das ideale Gas seine Gültigkeit behalten soll, muß man annehmen, daß die Gesetze idealer Gase bei niedrigen Temperaturen von den klassischen abweichen. Die Ursache dieser Entartung ist in einer Quantelung der Molekularbewegungen zu suchen. Bei allen Theorien der Entartung werden immer mehr oder weniger willkürliche Annahmen über das statistische Verhalten der Moleküle, oder über ihre Quantelung gemacht. In der vorliegenden Arbeit wird nur die von Pauli zuerst ausgesprochene und auf zahlreiche spektroskopische Tatsachen begründete Annahme benutzt, daß in einem System nie zwei gleichwertige Elemente vorkommen können, deren Quantenzahlen vollständig übereinstimmen. Mit dieser Hypothese werden die Zustandsgleichung und die innere Energie des idealen Gases abgeleitet; der Entropiewert für große Temperaturen stimmt mit dem Stern-Tetrodeschen überein.

In der klassischen Thermodynamik wird die Molekularwärme (bei konstantem Volumen)

$$c = \tfrac{3}{2} k T \tag{1}$$

gesetzt. Will man aber den Nernstschen Wärmesatz auch auf das ideale Gas anwenden können, so muß man (1) bloß als eine Näherung für große Temperaturen ansehen, da c im Limes für $T = 0$ verschwinden muß. Man ist deshalb genötigt, anzunehmen, daß die Bewegung der Moleküle idealer Gase gequantelt sei; diese Quantelung äußert sich bei niedrigen Temperaturen durch gewisse Entartungserscheinungen, so daß sowohl die spezifische Wärme als auch die Zustandsgleichung von ihren klassischen Ausdrücken abweichen werden.

Zweck der vorliegenden Arbeit ist, eine Methode für die Quantelung des idealen Gases darzustellen, welche nach unserem Erachten möglichst unabhängig von willkürlichen Annahmen über das statistische Verhalten der Gasmoleküle ist.

In neuerer Zeit wurden zahlreiche Versuche gemacht, die Zustandsgleichung idealer Gase festzustellen [2]). Die Zustandsgleichungen der verschiedenen Verfasser und unsere unterscheiden sich voneinander und

---

[1]) Vgl. die vorläufige Mitteilung, Lincei Rend. (6) **3**, 145, 1926.

[2]) Vgl. z. B. A. Einstein, Berl. Ber. 1924, S. 261; 1925, S. 318; M. Planck, ebenda 1925, S. 49. Unsere Methode ist der Einsteinschen insofern verwandt, als die Annahme der statistischen Unabhängigkeit der Moleküle bei beiden Methoden verlassen wird, obgleich die Art der Abhängigkeit bei uns ganz anders ist wie bei Einstein, und das Endergebnis für die Abweichungen von der klassischen Zustandsgleichung sogar entgegengesetzt gefunden wird.

von der klassischen Zustandsgleichung $pV = NkT$ durch Glieder, welche
nur bei sehr niedrigen Temperaturen und großen Drucken beträchtlich
werden; leider sind die Abweichungen der realen Gase von den idealen
gerade unter diesen Umständen am größten, so daß die an sich gar nicht
unbedeutenden Entartungserscheinungen bisher nicht beobachtet werden
konnten. Es ist jedenfalls nicht unmöglich, daß eine tiefere Kenntnis
der Zustandsgleichungen der Gase gestatten wird, die Entartung von den
übrigen Abweichungen von der Gleichung $pV = NkT$ zu trennen, so
daß eine experimentelle Entscheidung zwischen den verschiedenen Ent-
artungstheorien möglich wird.

Um die Quantenregeln auf die Bewegung der Moleküle unseres
idealen Gases anwenden zu können, kann man verschiedenartig verfahren;
das Endergebnis bleibt jedoch immer dasselbe. Z. B. können wir uns
die Moleküle in ein parallelepipedisches Gefäß mit elastisch reflektierenden
Wänden eingeschlossen denken; dadurch wird die Bewegung des zwischen
den Wänden hin und her fliegenden Moleküls bedingt periodisch und
kann deshalb quantisiert werden; allgemeiner kann man sich die Moleküle
in ein derartiges äußeres Kraftfeld eingesetzt denken, daß ihre Bewegung
bedingt periodisch werde; die Annahme, daß das Gas ideal ist, erlaubt
uns, die mechanischen Wirkungen der Moleküle aufeinander zu vernach-
lässigen, so daß ihre mechanische Bewegung sich nur unter dem Einfluß
der äußeren Kraft vollzieht. Es ist jedoch ersichtlich, daß die unter der
Annahme der vollständigen Unabhängigkeit der Moleküle voneinander
ausgeführte Quantelung der Molekularbewegungen nicht hinreichend ist,
uns Rechenschaft von der erwarteten Entartung zu geben. Das erkennt
man am besten an dem Beispiel der in einem Gefäß eingeschlossenen
Moleküle dadurch, daß, wenn die linearen Dimensionen des Gefäßes wachsen,
die Energiewerte der Quantenzustände jedes einzelnen Moleküls immer
dichter werden, so daß für Gefäße makroskopischer Dimensionen bereits
jeder Einfluß der Diskontinuität der Energiewerte praktisch verschwindet.
Dieser Einfluß hängt außerdem von dem Volumen des Gefäßes ab, auch
wenn die Zahl der im Gefäß enthaltenen Moleküle so gewählt wird, daß
die Dichte konstant bleibt.

Durch eine quantitative Berechnung dieses Sachverhaltes[1]) kann
man sich überzeugen, daß man nur dann eine Entartung der erwarteten
Größenordnung erhält, wenn man das Gefäß so klein wählt, daß es im
Mittel nur noch ein Molekül enthält.

---

[1]) E. Fermi, Nuovo Cim. 1, 145, 1924.

904   E. Fermi,

Wir sprechen deshalb die Vermutung aus, daß zur Quantelung idealer Gase eine Zusatzregel zu den Sommerfeldschen Quantenbedingungen nötig sei.

Nun wurde kürzlich von W. Pauli[1]), im Anschluß an eine Arbeit von E. C. Stoner[2]), die Regel aufgestellt, daß, wenn in einem Atom ein Elektron vorhanden ist, dessen Quantenzahlen (die magnetischen Quantenzahlen eingeschlossen) bestimmte Werte haben, so kann im Atom kein weiteres Elektron existieren, dessen Bahn durch dieselben Zahlen charakterisiert ist. Mit anderen Worten ist eine Quantenbahn (in einem äußeren magnetischen Felde) bereits von einem einzigen Elektron vollständig besetzt.

Da diese Paulische Regel sich in der Deutung spektroskopischer Tatsachen als äußerst fruchtbar erwiesen hat[3]), wollen wir versuchen, ob sie nicht etwa auch für das Problem der Quantelung idealer Gase brauchbar sei.

Wir werden zeigen, daß dies in der Tat der Fall ist, und daß die Anwendung der Paulischen Regel uns erlaubt, eine vollständig konsequente Theorie der Entartung idealer Gase darzustellen.

Wir werden also im folgenden annehmen, daß höchstens ein Molekül mit vorgegebenen Quantenzahlen in unserem Gase vorhanden sein kann; als Quantenzahlen kommen dabei nicht nur die Quantenzahlen in Betracht, welche die inneren Bewegungen des Moleküls, sondern auch die Zahlen, welche seine Translationsbewegung bestimmen.

Zuerst müssen wir unsere Moleküle in ein passendes äußeres Kraftfeld einsetzen, so daß ihre Bewegung bedingt periodisch werde. Das kann auf unendlich viele Weisen geschehen; da aber das Resultat von der Wahl des Kraftfeldes nicht abhängt, wollen wir die Moleküle einer zentralen elastischen Anziehung nach einem festen Punkte O (Koordinatensprung) unterwerfen, so daß jedes Molekül einen harmonischen Oszillator bilden wird. Diese Zentralkraft wird unsere Gasmenge in der Umgebung von O halten; die Gasdichte wird mit der Entfernung von O abnehmen und für unendliche Entfernung verschwinden. Sei ν die Eigenfrequenz der Oszillatoren, dann ist die auf die Moleküle wirkende Kraft durch

$$4\pi^2\nu^2 m r$$

gegeben, wo m die Masse der Moleküle und r ihre Entfernung von O darstellt. Die potentielle Energie der Anziehungskraft ist dann

$$u = 2\pi^2\nu^2 m r^2. \tag{1}$$

[1]) W. Pauli jr., ZS. f. Phys. **31**, 765, 1925.
[2]) E. C. Stoner, Phil. Mag. **48**, 719, 1924.
[3]) Vgl. z. B. F. Hund, ZS. f. Phys. **33**, 345, 1925.

Die Quantenzahlen des von einem Molekül gebildeten Oszillators seien s_1, s_2, s_3. Zur Charakterisierung des Moleküls sind eigentlich diese Quantenzahlen nicht hinreichend, denn dazu müßte man auch die Quantenzahlen der inneren Bewegungen angeben. Wir wollen uns aber auf einatomige Moleküle beschränken, und weiter wollen wir annehmen, daß alle in unserem Gase vorkommenden Moleküle sich im Grundzustand befinden, und daß dieser Zustand einfach (magnetisch unzerlegbar) ist. Dann brauchen wir uns um die inneren Bewegungen nicht zu kümmern, und wir können unsere Moleküle einfach als Massenpunkte ansehen. Die Paulische Regel lautet deshalb für unseren Fall: Es kann in der ganzen Gasmenge höchstens ein Molekül mit vorgegebenen Quantenzahlen s_1, s_2, s_3 vorhanden sein.

Die totale Energie dieses Moleküls wird durch

$$w = h\nu(s_1 + s_2 + s_3) = h\nu s \qquad (2)$$

gegeben. Die totale Energie kann deshalb ein beliebiges ganzzahliges Vielfaches von $h\nu$ sein; der Wert $s\,h\nu$ kann jedoch auf viele Weisen realisiert werden. Jede Realisierungsmöglichkeit entspricht einer Lösung der Gleichung

$$s = s_1 + s_2 + s_3, \qquad (3)$$

wo s_1, s_2, s_3 die Werte 0, 1, 2, 3, ... annehmen können. Bekanntlich hat Gleichung (3)

$$Q_s = \frac{(s+1)(s+2)}{2} \qquad (4)$$

Lösungen. Die Energie Null kann deshalb nur auf eine einzige Art realisiert werden, die Energie $h\nu$ auf drei, die Energie $2\,h\nu$ auf sechs usw. Ein Molekül mit der Energie $s\,h\nu$ werden wir einfach ein „s"-Molekül nennen.

Nach unseren Annahmen können nun in der ganzen Gasmenge höchstens Q_s „s"-Moleküle vorkommen; also höchstens ein Molekül mit der Energie Null, höchstens drei Moleküle mit der Energie $h\nu$, höchstens sechs mit der Energie $2\,h\nu$ usw.

Um die Folgen dieses Tatbestandes klar übersehen zu können, wollen wir den extremen Fall betrachten, daß die absolute Temperatur unseres Gases Null sei. Sei N die Zahl der Moleküle. Beim absoluten Nullpunkt muß sich das Gas in dem Zustand kleinster Energie befinden. Wäre nun keine Einschränkung für die Zahl der Moleküle einer gegebenen Energie vorhanden, so würde sich jedes Molekül im Zustand der Energie Null ($s_1 = s_2 = s_3 = 0$) befinden. Nach dem Vorhergehenden kann aber höchstens ein Molekül mit der Energie Null vorkommen; wäre deshalb

$N = 1$, so würde das einzige Molekül beim absoluten Nullpunkt den Zustand der Energie Null besetzen; wäre $N = 4$, so würde ein Molekül den Zustand der Energie Null, die drei übrigen die drei Plätze mit der Energie $h\nu$ besetzen; wäre $N = 10$, so würde sich ein Molekül am Platze mit der Energie Null befinden, drei andere an den drei Plätzen mit der Energie $h\nu$, und die sechs übrigen in den sechs Plätzen mit der Energie $2 h\nu$ usw.

Beim absoluten Nullpunkt zeigen deshalb die Moleküle unseres Gases eine Art schalenförmigen Aufbau, der eine gewisse Analogie zur schalenartigen Anordnung der Elektronen in einem Atom mit mehreren Elektronen aufweist.

Wir wollen jetzt untersuchen, wie sich eine gewisse Energiemenge

$$W = E h\nu \tag{5}$$

($E =$ ganze Zahl) zwischen unseren N Molekülen verteilt.

Sei N_s die Zahl der Moleküle, die sich in einem Zustand mit der Energie $s h\nu$ befinden. Nach unseren Annahmen ist

$$N_s \leq Q_s. \tag{6}$$

Man hat weiter die Gleichungen

$$\sum N_s = N, \tag{7}$$

$$\sum s N_s = E, \tag{8}$$

welche ausdrücken, daß die Gesamtzahl bzw. die Gesamtenergie der Moleküle gleich N bzw. $E h\nu$ ist.

Jetzt wollen wir die Zahl P solcher Anordnungen unserer N Moleküle berechnen, daß sich N_0 auf Plätzen mit der Energie Null, N_1 auf Plätzen mit der Energie $h\nu$, N_2 auf Plätzen mit der Energie $2 h\nu$ usw. befinden. Zwei Anordnungen sollen dabei als gleich angesehen werden, wenn die von den Molekülen besetzten Plätze dieselben sind; zwei Anordnungen, welche sich nur durch eine Permutation der Moleküle auf ihren Plätzen unterscheiden, sind deshalb als eine gleiche Anordnung anzusehen. Würde man zwei solche Anordnungen als verschieden ansehen, so würde man P mit der Konstante $N!$ multiplizieren müssen; man könnte aber leicht einsehen, daß dies auf das folgende keinen Einfluß haben würde. Im oben erklärten Sinne ist die Zahl der Anordnungen von N_s Molekülen auf den Q_s Plätzen der Energie $s h\nu$ durch

$$\binom{Q_s}{N_s}$$

gegeben. Wir finden deshalb für P den Ausdruck

$$P = \left(\frac{Q_0}{N_0}\right)\left(\frac{Q_1}{N_1}\right)\left(\frac{Q_2}{N_2}\right) \cdots = \prod \left(\frac{Q_s}{N_s}\right). \qquad (9)$$

Man bekommt die wahrscheinlichsten Werte der N_s, indem man das Maximum von P mit den Einschränkungen (7) und (8) sucht. Durch Anwendung des Stirlingschen Satzes kann man, mit für unseren Fall genügender Annäherung, schreiben:

$$\log P = \sum \log \left(\frac{Q_s}{N_s}\right) = - \sum \left(N_s \log \frac{N_s^!}{Q_s - N_s} + Q_s \log \frac{Q_s - N_s}{Q_s}\right). \quad (10)$$

Wir suchen also die Werte der N_s, welche (7) und (8) genügen, und für welche $\log P$ ein Maximum wird. Man findet:

$$\alpha e^{-\beta s} = \frac{N_s}{Q_s - N_s},$$

wo α und β Konstante darstellen. Die vorige Gleichung gibt uns:

$$N_s = Q_s \frac{\alpha e^{-\beta s}}{1 + \alpha e^{-\beta s}}. \qquad (11)$$

Die Werte von α und β können durch die Gleichung (7) und (8) bestimmt werden, oder umgekehrt kann man α und β als gegeben ansehen; dann bestimmen (7) und (8) die Gesamtzahl und die Gesamtenergie unserer Moleküle. Wir finden nämlich

$$\left. \begin{array}{l} N = \displaystyle\sum_0^\infty Q_s \frac{\alpha e^{-\beta s}}{1 + \alpha e^{-\beta s}}, \\[2mm] \dfrac{W}{h\nu} = E = \displaystyle\sum_0^\infty s\, Q_s \frac{\alpha e^{-\beta s}}{1 + \alpha e^{-\beta s}}, \end{array} \right\} \qquad (12)$$

Die absolute Temperatur T des Gases ist eine Funktion von N und E oder auch von α und β. Diese Funktion kann nach zwei Methoden bestimmt werden, welche jedoch zum selben Resultat führen. Man könnte z. B. nach dem Boltzmannschen Prinzip die Entropie

$$S = k \log P$$

setzen und dann die Temperatur nach der Formel

$$T = \frac{dW}{dS}$$

berechnen. Diese Methode hat jedoch, wie alle auf dem Boltzmannschen Prinzip beruhenden Methoden, den Nachteil, daß man für ihre Anwendung einen mehr oder weniger willkürlichen Ansatz für die Zustandswahrscheinlichkeit braucht. Wir ziehen deshalb vor, folgendermaßen zu verfahren: Beachten wir, daß die Dichte unseres Gases eine Funktion der Ent-

fernung ist, welche für unendliche Entfernung verschwindet. Für unendlich großes r werden deshalb auch die Entartungserscheinungen auf-
hören, und die Statistik unseres Gases in die klassische übergehen.
Insbesondere muß für $r = \infty$ die mittlere kinetische Energie der
Moleküle $3\,k\,T/2$ werden, und ihre Geschwindigkeitsverteilung in die
Maxwellsche übergehen. Wir können also die Temperatur aus der
Geschwindigkeitsverteilung in dem Gebiet unendlich kleiner Dichte bestimmen; und da die ganze Gasmenge auf konstanter Temperatur ist,
werden wir zugleich die Temperatur auch für die Gebiete hoher Dichte
kennen. Zu dieser Bestimmung werden wir uns sozusagen eines Gas-
thermometers mit einem unendlich verdünnten idealen Gase bedienen.

Zuerst müssen wir die Dichte der Moleküle mit einer kinetischen
Energie zwischen L und $L + d\,L$ in der Entfernung r berechnen. Die
totale Energie dieser Moleküle wird nach (1) zwischen

$$L + 2\,\pi^2\,v^2\,m\,r^2 \quad \text{und} \quad L + 2\,\pi^2\,v^2\,m\,r^2 + d\,L$$

liegen. Nun ist die totale Energie eines Moleküls gleich $s\,h\,\nu$. Für
unsere Moleküle muß s deshalb zwischen s und $s + d\,s$ liegen, wo

$$s = \frac{L}{h\,\nu} + \frac{2\,\pi^2\,v\,m}{h}r^2, \quad d\,s = \frac{d\,L}{h\,\nu}. \tag{13}$$

Betrachten wir jetzt ein Molekül, dessen Bewegung durch die Quantenzahlen s_1, s_2, s_3 charakterisiert ist. Seine Koordinaten x, y, z sind durch

$$\left.\begin{array}{l} x = \sqrt{H s_1}\,\cos\,(2\,\pi\,\nu\,t - \alpha_1), \quad y = \sqrt{H s_2}\,\cos\,(2\,\pi\,\nu\,t - \alpha_2), \\ z = \sqrt{H s_3}\,\cos\,(2\,\pi\,\nu\,t - \alpha_3) \end{array}\right\} \tag{14}$$

als Funktionen der Zeit gegeben. Dabei ist

$$H = \frac{h}{2\,\pi^2\,v\,m} \tag{15}$$

gesetzt worden; α_1, α_2 und α_3 bedeuten Phasenkonstanten, welche mit
gleicher Wahrscheinlichkeit jedes beliebige Wertesystem annehmen können.
Hieraus und aus den Gleichungen (14) folgt, daß $|x| \leqslant \sqrt{H s_1}$,
$|y| \leqslant \sqrt{H s_2}$, $|z| \leqslant \sqrt{H s_3}$, und daß die Wahrscheinlichkeit, daß x, y, z
zwischen den Grenzen x und $x + d\,x$, y und $y + d\,y$, z und $z + d\,z$ liegen,
folgenden Ausdruck hat:

$$\frac{d\,x\,d\,y\,d\,z}{\pi^3\,\sqrt{(H s_1 - x^2)(H s_2 - y^2)(H s_3 - z^2)}}.$$

Wenn wir nicht die einzelnen Werte von s_1, s_2, s_3, sondern nur
ihre Summe kennen, so ist unsere Wahrscheinlichkeit durch

$$\frac{1}{Q_s}\frac{d\,x\,d\,y\,d\,z}{\pi^3}\sum \frac{1}{\sqrt{(H s_1 - x^2)(H s_2 - y^2)(H s_3 - z^2)}} \tag{16}$$

ausgedrückt; die Summe ist auf alle ganzzahligen Lösungen der Gleichung (3) zu erstrecken, die den Ungleichungen

$$H s_1 \geqq x^2, \quad H s_2 \geqq y^2, \quad H s_3 \geqq z^2$$

genügen. Wenn wir die Wahrscheinlichkeit (16) mit der Anzahl N_s der „s"-Moleküle multiplizieren, so bekommen wir die Zahl der „s"-Moleküle, die im Volumenelement $dx\,dy\,dz$ enthalten sind. Unter Berücksichtigung von (11) finden wir also, daß die Dichte der „s"-Moleküle am Orte x, y, z durch

$$n_s = \frac{\alpha\,e^{-\beta s}}{1 + \alpha\,e^{-\beta s}} \frac{1}{\pi^3} \sum \frac{1}{\sqrt{(H s_1 - x^2)(H s_2 - y^2)(H s_3 - z^2)}}$$

gegeben ist. Für hinreichend großes s kann man die Summe durch ein zweifaches Integral ersetzen; nach Ausführung der Integrationen finden wir

$$n_s = \frac{2}{\pi^2 H^2} \frac{\alpha\,e^{-\beta s}}{1 + \alpha\,e^{-\beta s}} \sqrt{H s - r^2}.$$

Mit Benutzung von (13) und (15) finden wir jetzt, daß die Dichte der Moleküle mit einer kinetischen Energie zwischen L und $L + dL$ am Orte x, y, z folgenden Ausdruck hat:

$$n(L)\,dL = n_s\,ds = \frac{2\pi(2m)^{3/2}}{h^3} \sqrt{L}\,dL \frac{\alpha\,e^{-\frac{2\pi^2 v m \beta r^2}{h}} e^{-\frac{\beta L}{h v}}}{1 + \alpha\,e^{-\frac{2\pi^2 v m \beta r^2}{h}} e^{-\frac{\beta L}{h v}}}. \quad (17)$$

Diese Formel muß mit dem klassischen Ausdruck des Maxwellschen Verteilungsgesetzes verglichen werden:

$$n^*(L)\,dL = K \sqrt{L}\,dL\,e^{-L/kT}. \quad (17')$$

Man sieht dann, daß im Limes für $r = \infty$ (17) in (17') übergeht, wenn man nur

$$\beta = \frac{h v}{k T} \quad (18)$$

setzt. Jetzt kann (17) folgendermaßen geschrieben werden:

$$n(L)\,dL = \frac{(2\pi)(2m)^{3/2}}{h^3} \sqrt{L}\,dL \frac{A\,e^{-L/kT}}{1 + A\,e^{-L/kT}}, \quad (19)$$

wo

$$A = \alpha\,e^{-\frac{2\pi^2 v^2 m r^2}{kT}} \quad (20)$$

Die Gesamtdichte der Moleküle in der Entfernung r wird jetzt

$$n = \int_0^\infty n(L)\,dL = \frac{(2\pi m k T)^{3/2}}{h^3} F(A), \quad (21)$$

wo gesetzt worden ist:

$$F(A) = \frac{2}{\sqrt{\pi}} \int_0^\infty \frac{A\sqrt{x}\,e^{-x}\,dx}{1+A\,e^{-x}} \,. \tag{22}$$

Die mittlere kinetische Energie der Moleküle in der Entfernung r ist

$$\bar{L} = \frac{1}{n} \int_0^\infty L\,n(L)\,dL = \frac{3}{2}\,kT\,\frac{G(A)}{F(A)}, \tag{23}$$

wo

$$G(A) = \frac{4}{3\sqrt{\pi}} \int_0^\infty \frac{A\,x^{3/2}\,e^{-x}\,dx}{1+A\,e^{-x}} \,. \tag{24}$$

Mittels (21) kann man A als Funktion von Dichte und Temperatur bestimmen; wenn man den gefundenen Wert in (19) und (23) einsetzt, so bekommt man die Geschwindigkeitsverteilung und die mittlere kinetische Energie der Moleküle als Funktion von Dichte und Temperatur.

Zur Aufstellung der Zustandsgleichung wenden wir den Virialsatz an. Nach diesem ist der Druck durch

$$p = \frac{2}{3}\,n\,\bar{L} = n\,kT\,\frac{G(A)}{F(A)} \tag{25}$$

gegeben; der Wert von A ist wieder aus (21) als Funktion von Dichte und Temperatur zu entnehmen.

Ehe wir weitergehen, wollen wir einige mathematische Eigenschaften der eingeführten Funktionen $F(A)$ und $G(A)$ darstellen.

Für $A \leqq 1$ kann man beide Funktionen durch die konvergierenden Reihen

$$\left.\begin{aligned} F(A) &= A - \frac{A^2}{2^{3/2}} + \frac{A^3}{3^{3/2}} - \cdots, \\ G(A) &= A - \frac{A^2}{2^{5/2}} + \frac{A^3}{3^{3/2}} - \cdots \end{aligned}\right\} \tag{26}$$

darstellen. Für großes A hat man die asymptotischen Ausdrücke

$$\left.\begin{aligned} F(A) &= \frac{4}{3\sqrt{\pi}}\,(\log A)^{3/2}\left[1 + \frac{\pi^2}{8\,(\log A)^2} + \cdots\right], \\ G(A) &= \frac{8}{15\sqrt{\pi}}\,(\log A)^{5/2}\left[1 + \frac{5\,\pi^2}{8\,(\log A)^2} + \cdots\right]. \end{aligned}\right\} \tag{27}$$

Es gilt weiter die Beziehung

$$\frac{d\,G(A)}{F(A)} = d\log A. \tag{28}$$

Wir müssen noch eine andere, durch die Beziehungen

$$P(\Theta) = \Theta \frac{G(A)}{F(A)}, \quad F(A) = \frac{1}{\Theta^{3/2}} \qquad (29)$$

definierte Funktion $P(\Theta)$ einführen. Für sehr großes bzw. sehr kleines Θ kann $P(\Theta)$ mit den Näherungsformeln

$$P(\Theta) = \Theta \left\{ 1 + \frac{1}{2^{5/2}\,\Theta^{3/2}} + \cdots \right\}$$

bzw.

$$P(\Theta) = \frac{3^{2/3}\,\pi^{1/3}}{5 \cdot 2^{1/3}} \left\{ 1 + \frac{5 \cdot 2^{2/3}\,\pi^{4/3}}{3^{7/3}}\,\Theta^2 + \cdots \right\} \qquad (30)$$

berechnet werden. Mit Benutzung von (29), (28), (27) erkennt man weiter, daß

$$\int_0^\Theta \frac{d\,P(\Theta)}{\Theta} = \frac{5}{3}\,\frac{G(A)}{F(A)} - \frac{2}{3}\log A. \qquad (31)$$

Wir sind jetzt imstande, aus der Zustandsgleichung (25) und der Gleichung (28) den Parameter A zu eliminieren, und wir finden den Druck und die mittlere kinetische Energie der Moleküle als explizite Funktionen von Dichte und Temperatur:

$$p = \frac{h^2\,n^{5/3}}{2\,\pi\,m}\,P\!\left(\frac{2\,\pi\,m\,k\,T}{h^2\,n^{2/3}}\right), \qquad (32)$$

$$\overline{L} = \frac{3}{2}\,\frac{h^2\,n^{2/3}}{2\,\pi\,m}\,P\!\left(\frac{2\,\pi\,m\,k\,T}{h^2\,n^{2/3}}\right). \qquad (33)$$

Im Grenzfall schwacher Entartung (T groß und n klein) nimmt die Zustandsgleichung folgende Form an:

$$p = n\,k\,T \left\{ 1 + \frac{1}{16}\,\frac{h^3\,n}{(\pi\,m\,k\,T)^{3/2}} + \cdots \right\}. \qquad (34)$$

Der Druck ist also größer als nach der klassischen Zustandsgleichung ($p = n\,k\,T$). Für ein ideales Gas mit dem Atomgewicht von Helium, bei $T = 5^0$ und einem Druck von 10 Atm. beträgt der Unterschied etwa 15 Proz.

Im Grenzfall großer Entartung nehmen (32) und (33) die Form

$$p = \frac{1}{20}\left(\frac{6}{\pi}\right)^{2/3}\frac{h^2\,n^{5/3}}{m} + \frac{2^{4/3}\,\pi^{3/3}}{3^{5/3}}\,\frac{m\,n^{1/3}\,k^3\,T^2}{h^2} + \cdots \qquad (35)$$

$$\overline{L} = \frac{3}{40}\left(\frac{6}{\pi}\right)^{2/3}\frac{h^2\,n^{2/3}}{m} + \frac{2^{1/3}\,\pi^{8/3}}{3^{2/3}}\,\frac{m\,k^2\,T^2}{h^2\,n^{2/3}} + \cdots \qquad (36)$$

an. Man erkennt hieraus, daß die Entartung einen Nullpunktsdruck und eine Nullpunktsenergie zur Folge hat.

Aus (36) kann man auch die spezifische Wärme für tiefe Temperaturen berechnen. Man findet

$$c_v = \frac{d\overline{L}}{dT} = \frac{2^{4/3}\,\pi^{8/3}}{3^{2/3}}\,\frac{m\,k^2\,T}{h^2\,n^{2/3}} + \cdots \tag{37}$$

Man erkennt, daß die spezifische Wärme beim absoluten Nullpunkt verschwindet, und zwar daß sie für tiefe Temperaturen der absoluten Temperatur proportional ist.

Zuletzt wollen wir zeigen, daß unsere Theorie zum Stern-Tetrodeschen Wert für die absolute Entropie des Gases führt. Durch Anwendung von (33) findet man in der Tat

$$S = n\int_0^T \frac{d\overline{L}}{T} = \frac{3}{2}\,n\,k\int_0^\Theta \frac{P'(\Theta)\,d\Theta}{\Theta}.$$

(31) gibt uns jetzt

$$S = n\,k\left\{\frac{5}{2}\,\frac{G(A)}{F(A)} - \log A\right\}, \tag{38}$$

wo der Wert von A wieder aus (21) zu entnehmen ist. Für hohe Temperaturen finden wir deshalb mit Anwendung von (26)

$$A = \frac{n\,h^3}{(2\,\pi\,m\,k\,T)^{3/2}}, \qquad \frac{G(A)}{F(A)} = 1.$$

(38) gibt uns dann

$$\begin{aligned}
S &= n\,k\left\{\log\frac{(2\,\pi\,m\,k\,T)^{3/2}}{n\,h^3} + \frac{5}{2}\right\} \\
&= n\,k\left\{\frac{3}{2}\log T - \log n + \log\frac{(2\,\pi\,m)^{3/2}\,k^{3/2}\,e^{5/2}}{h^3}\right\},
\end{aligned}$$

was mit dem Entropiewert von Stern und Tetrode übereinstimmt.

Acknowledgments

The editors wish to thank the Institute of Electrical and Electronics Engineers (IEEE) for kind permission to reproduce in this book two articles that have appeared in IEEE publications.

The editors are also grateful to:

Franco Angotti – University of Florence
Gorgio Baccarani – University of Bologna
Ermanno Cardelli – IEEE Italy Section Chairman
Roberto Casalbuoni – University of Florence
Alessandro Cidronali – University of Florence
Robert Colburn – IEEE History Center, Research Coordinator
Vaishali Damle – Proceedings of the IEEE Managing Editor
Luigi Dei – University of Florence, Rector
Enrico Del Re – University of Florence
Daniele Dominici – University of Florence
Mahta Moghaddam – IEEE Antennas and Propagation Magazine, Editor-in-Chief
Massimiliano Pieraccini – University of Florence
Antonio Savini – Universiy of Pavia
Stefano Selleri – University of Florence
W. Ross Stone – IEEE Antennas and Propagation Magazine, Former Editor-in-Chief
Alberto Tesi – University of Florence, Former Rector
Piero Tortoli – University of Florence

The Authors

Giorgio Baccarani
Alma Mater Studiorum, University of Bologna
Via Zamboni 33
Bologna, Italy
giorgio.baccarani@unibo.it

Ermanno Cardelli
Department of Engineering, University of Perugia
Via G. Duranti 67
Perugia, Italy
ermanno.cardelli@unipg.it

Gianfranco Manes
Department of Information Engineering, University of Florence
Via di S. Marta 3
Florence, Italy
gianfranco.manes@unifi.it

Giuseppe Pelosi, Massimiliano Pieraccini, Stefano Selleri
Department of Information Engineering, University of Florence
Via di S. Marta 3
Florence, Italy
[giuseppe.pelosi,massimiliano.pieraccini,stefano.selleri]@unifi.it

TITLES PUBLISHED

1. Casalbuoni R., Frosali G., Pelosi G. (editors), *Enrico Fermi a Firenze. Le «Lezioni di Meccanica Razionale» al biennio propedeutico agli studi di Ingegneria: 1924-1926* [*Enrico Fermi in Florence. The lessons of «Mechanics» at the two-year preparatory courses of Engineering: 1924-1926*]
2. Manes G., Pelosi G. (editors), *Enrico Fermi's IEEE Milestone in Florence. For his Major Contribution to Semiconductor Statistics, 1924-1926*